Animal portraits

Richard Lydekker

Alpha Editions

This edition published in 2024

ISBN : 9789367244814

Design and Setting By
Alpha Editions
www.alphaedis.com
Email - info@alphaedis.com

As per information held with us this book is in Public Domain.
This book is a reproduction of an important historical work. Alpha Editions uses the best technology to reproduce historical work in the same manner it was first published to preserve its original nature. Any marks or number seen are left intentionally to preserve its true form.

Contents

FOREWORD	- 1 -
THE LION	- 3 -
THE MANCHURIAN TIGER	- 7 -
THE SEAL	- 9 -
THE ELK	- 11 -
THE CAPERCAILLIE	- 13 -
THE SQUIRREL	- 16 -
THE ROE-DEER	- 19 -
THE REINDEER	- 22 -
THE DEFASSA WATERBUCK	- 25 -
THE HARE	- 27 -
THE POLAR BEAR	- 30 -
THE MANDRILL	- 33 -
THE WOLF	- 36 -
THE WILD CAT	- 39 -
THE RED KANGAROO	- 42 -
THE BLUE ROLLER	- 45 -
THE BITTERN	- 48 -
THE KAFIR CROWNED CRANE	- 51 -
THE SILVER GULL	- 54 -
THE GREAT HORNED OWL	- 56 -
THE FLAMINGO	- 59 -
THE NILE CROCODILE	- 62 -
THE WHITE-HANDED GIBBON	- 65 -
THE ABYSSINIAN GREEN MONKEY	- 68 -
THE FOX	- 71 -

THE BROWN BEAR	- 74 -
THE PINE-MARTEN	- 77 -
THE LEOPARD	- 80 -
THE LYNX	- 83 -
THE INDIAN RHINOCEROS	- 86 -
THE BISON	- 89 -
THE GAZELLE	- 92 -
THE MOUFLON	- 95 -
THE RED DEER	- 98 -
THE BEAVER	- 101 -
THE MARMOT	- 104 -
THE HAMSTER	- 107 -
THE DUCKBILL OR PLATYPUS	- 110 -
THE SPINY ANT-EATER OR ECHIDNA	- 113 -
THE BLACK SWAN	- 116 -
THE BUSTARD	- 119 -
THE SOMALI OSTRICH	- 122 -
THE PIED HORNBILL	- 124 -
THE BLUE MACAW	- 127 -
THE MANDARIN DUCK	- 130 -
THE HERON	- 133 -
THE GREAT SPOTTED WOODPECKER	- 136 -
THE SENEGAL PARROT	- 139 -
THE GOLDFINCH	- 142 -
THE RAZORBILL OR AUK	- 145 -

FOREWORD

WILHELM KUHNERT is the greatest animal-painter of our day. There has never been a more excellent colourist or more skilful draughtsman of shape and action; and there is none with a more masterly touch or more unmistakable individuality. There is no misunderstanding his intention. The lifelike likeness of the animal set in its local atmosphere is his predominating endeavour. His range is of the widest, and his subjects all come alike to him; he has no preference for any particular group or family, but draws them all, vertebrate and invertebrate, with power, truth, and sympathy.

With a knowledge of the very soul of the animal such as few possess, his pictures are replete with insight into character and its vivid expression, and fascinate even those who may not adequately appreciate their wonderful accuracy. "We who have travelled," as Mr. G. J. Millais remarks, "do not need to be told that his studies from nature are correct. His lions, elephants, zebras, and antelopes are so real that we feel we are gazing at them on the plains of East Africa. The landscapes are simple but intense; sunlight is there, and the trees and grass are just those that grow in the habitat of these species. Kuhnert has, as it were, got inside the very skin of African life, and draws you insensibly within the charmed circle."

In a gallery his works at once arrest attention by their vigorous realism. There is life within the frame of whatever he paints. He is the Frans Hals of animal portraiture.

He was born at Oppeln in Silesia on the 28th of September 1865, and during his student days at the Academy of Berlin, was influentially advised to devote himself to animal painting, for which he had evidently a special gift. He began, however, as a portrait painter, and from his pictures, particularly those of African life, it is clear that in that branch of art he would have distinguished himself; but fortunately, he could not withstand his inclination. From a painter of portraits of men and women he developed into a portrait-painter of animals, finding his subjects alive by the countryside and in menageries and zoological gardens, and then seeking them farther afield in Africa and Southern Asia, where he worked assiduously in forest and jungle. Fifty of his characteristic studies are included in this book, and in them, as in all, the blending of the animal with the surroundings is remarkable, and the faithfulness with which the landscape, painted on the spot, has been rendered is apparent at a glance.

Search as we will we shall find nothing truer to nature than such triumphs of art as the Polar Bear amid the Arctic ice, the hairy Tigers by the snowy mountain lake, the slender Flamingoes in the evening landscape, the Silver

Gull sweeping above the ocean waves, the Black Swans as an idyll of the pool, the Capercaillie posturing in the morning light, and many other masterpieces herein.

This collection, like that of the Kuhnert Exhibition at the Fine Art Society's Gallery in Bond Street, is arranged purely on artistic lines; no attempt has been made to classify the subjects in zoological order. It is an album of animal portraiture as fully representative of the artist as possible. The pictures have been photographically reproduced in colours from the oil paintings, and the imitation is so exact that little of the charm of the originals has been lost.

THE LION

(*Felis leo*)

THE lion, undoubtedly, owes his title of "king of beasts" to the flowing mane with which his head and fore-quarters are adorned, as this confers upon him a dignity and grandeur of appearance entirely lacking in his maneless partner. Without his mane it is, indeed, more than questionable whether a lion would not be outclassed in style by a tiger. As it is, however, the lion, at all events so far as appearance goes, has an undisputed claim to his royal title; the magnificent mien of his draped head and shoulders, his bold, imperious eye, the powerful build of his lithe body, and his resounding roar of defiance, presenting the very ideal of supreme strength and sovereignty.

The tawny coat, which varies in tint from greyish or yellowish brown to yellow, is evidently intended to harmonise with the dry grasses and the yellow sand of the semi-desert tracts which form the favourite haunts of the lion; and this is confirmed by the fact that the newborn cubs are mottled with dark brown, thus indicating their descent from a species with a mottled or spotted coat adapted to a different environment. Indeed, Somali lions frequently retain traces of these spots, more especially in the female; and in German East Africa there exists a race of the lion in which both sexes are more or less fully and distinctly spotted.

In popular estimation the lion is inseparably connected with Africa, where it formerly ranged from one end of the continent to the other, although it has long since disappeared from most parts of Tunis and Algeria, as well as from Cape Colony. As a matter of fact, *Felis leo* is as much an Asiatic as an African animal; while in the time of Xenophon it was probably found so far west as Thrace and Macedonia. Earlier still, that is to say in prehistoric times, its range included the greater part of temperate Europe, not excepting the British Isles. Unlike the tiger, however, the lion never inhabited the countries to the east of the Bay of Bengal, nor penetrated to the swamps of Lower Bengal itself, which are unsuited to its habits. Even so late as the Mutiny, lions were to be met with over a considerable tract in central India; although they are nowadays restricted to that district of Kathiawar, known as the Gir, where they survive only by protection. In the valleys of the Euphrates and Tigris, as well as in parts of Persia, lions are still to be met with; but how numerous they are in these countries is difficult to ascertain.

In Africa lions appear to be most abundant in the British, German, and Portuguese eastern provinces, in some districts of which they seem bolder and more prone to attack human beings than in many other parts of the continent. In Somaliland, where they are smaller and greyer than usual, their number has been greatly reduced of late years. The handsomest lions of all are those with dark brown or black manes; but in some parts of the country, at all events, black-maned lions do not form a distinct race, as dark and light maned cubs may be found in the same litter.

Although the skulls of the two species present considerable differences, a lioness, apart from the absence of stripes, is not unlike a tiger in general appearance; and it has long been a question whether the lion or the tiger is the more powerful animal, although the balance of opinion seems to be in favour of the tiger's claim to superiority in this respect. Certain it is, that the lion is much the more noisy animal of the two; a tiger never roaring in the persistent manner characteristic of the lion. The impression caused by the lion's roar appears to depend greatly on the idiosyncrasy of the listener and the circumstances under which it is heard. Very noteworthy is the fact that the roar of the two species is essentially similar in character.

As a rule, lions, when too feeble to capture more active prey, turn into regular man-eaters much less frequently than tigers; but this may be in part explained by the bolder nature of many African tribes, as compared with the natives of a large extent of India; and when a lion makes himself obnoxious, such tribes have no hesitation in attacking and destroying the marauder.

Lions in Africa subsist, to a great extent, on the flesh of antelopes and zebras, or bontequaggas; generally stalking their game in parties of two or three, but one alone making the fatal spring. When stalking, a lion stretches out its body to the fullest extent, and crawls so close to the ground that even in low grass its presence is generally undetected till too late. Occasionally a party of lions combine their forces to pull down a large animal like a buffalo. Zebras can defend themselves only by kicking, but the gemsbok and the sable antelope will pin a lion with their horns, and sometimes come off victorious.

It has been very generally stated that lions are mainly, if not exclusively, monogamous, and that they mate for life. It has, however, been pointed out by Mr. Roosevelt that if this were really the case, they would almost always be found in pairs, that is to say, a lion and a lioness together. That they are thus found not infrequently may, indeed, be freely admitted; but, on the other hand, it is much commoner to come across a lioness and her cubs, an old lion with several lionesses and their young, a single lion or lioness, a couple of lions and lionesses, or, lastly, a small troop, which may be composed either solely of males or females, or of a mixture of the two sexes. "These facts," writes the great American hunter, "are not compatible with the romantic theory in question."

The cubs, generally two or three in number, come into the world, unlike kittens, with their eyes open, and are then about one-third the size of a cat. As already mentioned, they are heavily mottled with brown on a tawny ground, and it is very significant that these markings are to a great extent intermediate in character between the rosettes of leopards and jaguars and the stripes of the tiger. A peculiar feature in which the cubs of lions differ from those of leopards and jaguars, and thereby resemble those of tigers, is the presence of a white spot near the summit of the back of each ear. From these facts it has been inferred, in the first place, that the lion is most nearly related to the tiger, and, in the second place, that lions, tigers, leopards, and jaguars are all members of a single group. As regards the mutual relationships of these species, it is generally believed that spots represent an earlier and more primitive type of colouring than transverse stripes, and it is therefore inferred that the stripes of the tiger, which are very frequently partially split or double, have been derived from the fusion of leopard-like rosettes into transverse chains. As the self-coloured coat of the adult lion is evidently a modern feature, it seems clear that tigers and lions are to be regarded as the most specialised members of the whole group.

Before the investigations which led up to these modern advanced views had been undertaken, it was very generally believed, on account of its self-coloured tawny coat, that the American puma—locally known as the American lion—was one of the nearest relatives of *Felis leo*. If, however, beauty be but skin deep, colour is an even less deeply seated feature among animals; and, as the result of the study of the markings of young cubs of the puma, it seems certain that this species has acquired its uniform tawny livery quite independently of the lion. For newly born puma cubs exhibit a pattern of quadrangular blackish markings totally different in form and arrangement from those of young lions, tigers, leopards, or jaguars, and approximating in some degree to those of the smaller cats. Accordingly, in the opinion of the investigator to whom we are indebted for these very interesting views with regard to the inter-relationships of the various members of the feline tribe, it seems highly probable that the puma may be an overgrown self-coloured representative of the group of smaller cats typified by the ordinary domesticated species and its wild relatives.

A remarkable feature connected with the tuft at the tip of the tail of the lion is the frequent presence of a horny spur or claw, the function of which is still unknown, although it is certain that it is not employed, as was once thought to be the case, to goad the animal into fury when the tail is lashed against the flanks. It has been asserted that this spur is found only in the Indian lion; but this is as erroneous as the statement made by the same writer that it represents the last joint of the vertebræ of the tail to which the blood is unable to obtain access. A very similar structure exists in one member of the kangaroo tribe, known as the spur-tailed wallaby, in which, however, the spur is common to both sexes and quite constant in its development.

Menagerie lions, it may be mentioned, generally display a greater luxuriance and profusion of mane than their wild relatives; while it is in the former alone that any marked development of long hair on the under surface of the body is noticeable. The reason for this is, of course, too obvious to require explanation, more especially when it is borne in mind that lions inhabiting open plains with grass-jungle have larger manes than those which have to get their living in a country overgrown with thorn-bushes.

THE MANCHURIAN TIGER

(Felis tigris mongolica)

THE tiger is, and apparently always has been, an essentially Asiatic animal, although it enters south-eastern Europe in the Caucasus, whence its range extends eastwards through Persia, Afghanistan, and India, to Java and Sumatra, while northward it is found through China and Mongolia to Korea and Amurland. Very noteworthy is its absence from Ceylon, which seems to indicate that its original home was central Asia, and that it is a comparatively recent immigrant into southern Asia.

Four local races of the tiger are recognised by naturalists, namely, the typical Indian tiger, the Persian tiger (*F. tigris virgata*), the Javan tiger (*F. t. sondaica*), and the Manchurian or Siberian tiger (*F. t. mongolica* or *longipilis*). As regards splendour of coat, the finest of these races is the Manchurian tiger, which differs from the Indian race by the great length and woolliness of its winter coat and the larger extent of white on the face, under-parts, and the inner side of the limbs. It also appears to be a more stoutly built animal, and attains a very large size. Its skin commands a very high price. From both these races the Persian tiger, which is probably the one found in the Caucasus, differs by the copious fringe of long hair on the cheeks, throat, and under-parts; while it is also of inferior bodily size. The Javan and Sumatran tiger, which may be the same as the one inhabiting the Malay countries, differs from all the other three by the light areas on the head, body and limbs being of small extent, ill-defined, and dirty or buffish white in colour instead of pure white. In size it is always relatively small, and appears to be the smallest of all. The degree of development of the dark stripes appears to vary individually; but the Persian race seems to be the most fully striped of all, and shows in perfection the characteristic looping, or splitting of the stripes.

The striped coat of the tiger seems designed to break up the outline of the body, and thus to render the animal as inconspicuous as possible. That it is not, as was once thought, a special adaptation to match the surroundings of the animal in the jungles of Bengal, will be evident from what is stated above as to the distribution of the tiger and its comparatively recent entry into India.

Information is still required with regard to the habits of the Manchurian tiger, which has only of recent years been hunted by European sportsmen, or exhibited alive in zoological gardens. It appears, however, that these northern tigers prey on almost all the animals of their native country, from the largest to the smallest. In summer they will overpower and kill such fierce animals as bears, while in times of scarcity during the long winter they may be driven to prey on mice and rats. To capture their victims they frequently resort to the drinking-places of the latter, and in summer they likewise pay constant visits to the salt-pans where the wapiti and other deer come to lick the salt.

Each pair of tigers generally frequents the same lair, which in mountainous districts is concealed amid rocks, while in the plains the requisite shelter is afforded by reed-brakes. In winter the tigers may be completely snowed up in their retreats. As in India, the tigress is reported to separate herself from the tiger a short time before giving birth to her young; these being born in some dense thicket or rocky cleft, their number varying from two to four. At birth they are about the same size as lion-cubs, but are marked in the same fashion as their parents, although clothed in more woolly coats. In India it is generally stated that the female keeps apart from the male while the cubs are young, although both parents have been seen with their offspring. Like lions, tigers associate for the greater part of the year in pairs, and are strictly monogamous.

Whether old tigers in Siberia turn man-eaters, after the fashion of their brethren in India, does not appear to be known; but in much of their range it is probable that they do not resort to this mode of livelihood, on account of the sparseness of the population. It appears, however, that throughout northern China and Siberia the natives regard the tiger with much the same superstitious awe as do the natives of Hindustan. All the Siberian natives consider the tiger as a being of high nature, although the Tunguses look upon it as an evil spirit which has come under the influence of the Shamans, or sorcerers.

Fossil bones from the New Siberian Islands indicate that the Manchurian tiger once ranged beyond the Arctic Circle.

THE SEAL

(*Phoca vitulina*)

OF all mammals except whales, dolphins, and sea-cows, the true seals, of which the common British species is the typical representative, are those which have become most completely adapted to an aquatic existence. This is shown by the spindle-shaped form of the body, the total absence of external ears, the conversion of the fore-limbs into flippers, and the backward direction of the hind pair, which are also flipper-like, and lie parallel to the short tail to form an efficient rudder-like organ.

Seals are evidently descended from land Carnivora, and possibly from that long extinct group the Creodontia. When fully developed, their cheek-teeth consist of a single middle cusp, flanked by a smaller one in front and behind; and there are no teeth specially corresponding with the scissor-like pair characteristic of the modern land Carnivora. This fact, unless the teeth are degenerate, is all in favour of the direct descent of seals from creodonts.

From the true seals of the family *Phocidæ* the eared seals, or *Otariidæ*, differ by retaining small external ears, as well as by the fact that the hind-flippers are not permanently turned backwards; this latter feature being also distinctive of the walrus (*Odobænidæ*). In addition to being the most specialised of the aquatic Carnivora, the true seals are also the most widely distributed, inhabiting nearly all seas, and being likewise found in the Caspian Sea and in Lakes Aral and Baikal. Although seals go on land nightly to sleep, and likewise spend a considerable amount of time in the day on the shore or on ice, while the females give birth to their young on land, none of the *Phocidæ* regularly leave the water for a period of several weeks during the breeding-season, after the fashion of their eared cousins. Neither do any of them yield commercial sealskin, which is a product of certain members of the eared groups. Commercially, they are therefore valued only for their hides and oil, for the sake of which vast numbers are annually slaughtered.

Seals are adepts in swimming and diving, and have the power of closing their nostrils and the apertures of their ears while under water, although they are unable to remain beneath the surface for anything like so long as whales and dolphins. The food of the common seal, as well as most other kinds, consists chiefly of fishes, for capturing and holding which their sharply cusped teeth are admirably adapted. Graceful and active as are their movements in water, on land they are comparatively awkward and ungainly; progression being effected by means of the limbs, accompanied by sudden flexures of the body, so that it in some degree partakes of the nature of hopping. To land, seals shoot themselves out of the water by a strong and sudden movement of the hind-limbs. When once on land, they may remain there for days and even weeks together, till compelled by hunger to return to the sea.

The senses of the seal are highly developed. The eye, for instance, is full and globular, and thus specially adapted to catch every ray of light when the animal is in the water. It is noteworthy that seals can shed tears under the influence of excitement, and especially when in pain. And the idea of the ancients that these animals are attracted by music and singing appears to be founded on fact.

Their usual cry is a sharp bark, but when angered they give vent to dog-like snarlings. Seals always associate in parties, which may comprise hundreds of individuals. The young, which are beautiful little creatures, are tenderly and affectionately nurtured by their parents, who protect them from danger by every means in their power.

Young seals have a coat quite different from that of the adult; this baby coat in the case of the second British species, the grey seal (*Halichærus grypus*), being wholly white. In colour, the common seal varies greatly, the coat being sometimes yellowish and sometimes light grey with blackish mottlings and marblings. The so-called grey seal is generally much darker in colour, but is best distinguished from the common species by its greatly superior size and its relatively larger and blunter teeth.

All seals are polygamous, which leads to fights among the males for the possession of the females; but in other respects they are gentle and affectionate, easily tamed if taken young, and displaying great capacity for education. As a rule, only a single cub is produced at a birth, but there may be twins.

THE ELK

(Alces machlis)

ALTHOUGH the elk can claim an easy superiority in the matter of size over all the other members of the deer tribe, it certainly cannot be accorded a high position in the scale of beauty. For, truth to say, it is an ugly and ungainly creature, with disproportionately long legs, and huge head terminating in a broad, flabby, and almost trunk-like muzzle. By the sportsman, however, it is held in high estimation, owing to the magnificent trophies formed by its great spreading antlers, which in Alaskan specimens may have a span of as much as six feet. And when the build of the elk is considered in relation to its mode of life, we see that what appears ugly and ungainly to our eyes is merely adaptation to a particular mode of life. For in summer the elk spends much of its time wading belly-deep in marshes and lakes in search of the water-plants which form a large proportion of its food at this season; and in this pursuit its long limbs must obviously be of the greatest advantage, while the broad and mobile muzzle is specially well adapted for gathering in the floating leaves and stalks. Possibly the almost wholly hairy extremity of the muzzle is another adaptation to the same end. The elk typifies the wading type among mammals just as much as does the flamingo among birds.

Like many other mammals of northern Europe, the elk has a circumpolar distribution, although most Transatlantic naturalists regard its American representatives in the light of a distinct species rather than as local races.

In common with the brown bear, the elk attains its maximum stature in Alaska, where it towers to a height of close on seven feet at the shoulder. At one time an inhabitant of the British Isles, the elk is still found in many parts of Germany and Austria-Hungary, and is abundant in Scandinavia; from these countries its range extends eastward through Poland and Russia, and thence across the whole of Siberia. On the other side of Bering Strait it reappears in Alaska, whence it ranges through British Columbia to Maine and other parts of the United States. The differences between the elk (or moose, as it is there called) of the United States and the typical elk of Scandinavia are so slight that it requires an expert to distinguish between the two.

It is, however, very noteworthy that certain Scandinavian elk never develop the huge expansions, or "shovels," which form the most characteristic feature of the antlers of the species, but carry only five simple tines; and in east Siberia this simpler type of antler seems to be very prevalent.

A peculiar feature of the bull elk is the curious hairy appendage hanging from the throat, known to hunters as the "bell."

Elk are polygamous, like the majority of the deer tribe; and in the breeding season the two sexes learn each others' whereabouts by means of a loud "call" or bellowing, which in some districts, at any rate, appears to be uttered by males and females alike. The call can easily be imitated with the aid of a horn or trumpet, and by this means many a fine old bull is lured to his destruction. Elk are adepts in concealing themselves in the thickets to which they resort during the daytime.

In winter, when they are compelled to subsist on bark and twigs, especially those of the birch, these giant deer experience very hard times; and in North America a bull and two or three cows often form what is called a "yard" in the forest, by constantly trampling down the snow over a certain area, and thus keeping themselves from being snowed up. The female gives birth to one or two calves at a time, which are even more ungainly-looking than their parents.

It should be added, that in America the term "elk" is misapplied to the wapiti, while in Ceylon it is bestowed on the sambar deer.

THE CAPERCAILLIE

(Tetrao urogallus)

FOR the greater part of the year capercaillie, the largest representative of the grouse family, passes its time concealed in the depths of the forests, where it manages to find sufficient food even in the most severe winters; and it is only for a short period in spring that it makes its appearance, during the breeding-season, in the open. Its home is in the great forests of continental Europe and northern Asia, more especially those in which fir and pines predominate; abundant under-wood, which affords a good supply of berries, open glades, patches of sand, and pure water are, however, essential to the well-being of this magnificent bird. The capercaillie, or auerhahn as it is called in Austria and Germany, ranged in former days from the British Islands to the north-eastern portion of Turkestan, the Altai Mountains, and Lake Baikal; but by the middle of the seventeenth century it had already become scarce in Britain, where it became extinct a century later. In recent years the bird has, however, been reintroduced into Perthshire, Forfarshire, and a few other Scottish counties. In the Urals, north-eastern Siberia, and Kamchatka the typical capercaillie is represented by nearly allied species, or races. In Scandinavia the auerhahn ranges as far north as latitude 70°, but gradually becomes smaller and scarcer as the pine-woods tend to disappear; and it is the vast pine-forests of central Germany, Austria-Hungary, and Poland that form the great centre of its habitat.

Although its large size constitutes a sufficient means of recognising the cock capercaillie, which measures as much as thirty-five inches in length, from all other kinds of grouse, the species is distinguishable from its near relative, the blackcock, by the evenly rounded tail; while the blackish head and neck, with a patch of bare vermilion skin above each eye, the pale horn-colour of the beak, the green band across the breast, and the slaty brown back form other unmistakable characteristics. The female presents somewhat more resemblance to an overgrown greyhen (the female of the blackcock), but here again the rounded tail and superior size constitute decisive points of difference; while the general colour is more distinctly chestnut. Additional peculiarities of the hen capercaillie are to be found in the presence of a rufous patch at the base of the neck, and in the white tips to the black greater wing-coverts. In length the hen measures about 10 inches less than her partner; while her weight is only from 5 to 6 lb., against from 13 to 16 or 17 lb. in the cock.

The food of the cock capercaillie consists principally of the leaves and young shoots of the Scotch pine; and during the breeding-season these appear to form its sole nutriment. At other times he also consumes the leaves of deciduous trees, together with cranberries, whortleberries, juniper-berries, and grass, and sometimes a few insects or worms. On the other hand, the hens, which spend much more time on the ground than the cocks, and the young feed largely upon ants and other insects, worms, etc.; this being especially the case with the young birds.

The capercaillie is a polygamous bird, and in the spring—sometimes while the snow is still thick on the ground—each cock takes up a position on a tall pine, and commences his nuptial performance or *spel*, as it is called in Norway. This *spel* usually lasts from early dawn to sunrise, and is repeated from sunset till dark; but the time varies somewhat according to the state of the weather and the temperature. During this amatory performance the cock stretches out his neck, raises and spreads out his wings like a fan, ruffles his feathers, and assumes an appearance which has been compared to that of an angry turkey. A call, which has been expressed by the words *peller, peller, peller,* is uttered with continually increasing rapidity, and concludes with a gulp and an indrawing of the breath, when the bird throws up his head, closes his eyes, and appears to be in a paroxysm of passion.

At the sound of the cock's call all the hens in the neighbourhood rush to the place, uttering cries somewhat like the croak of a raven; and when they have assembled on the ground below the cock descends from his perch to join the party.

In Scotland, capercaillie-shooting takes place from 10th August till 20th December; but in some parts of the Continent, as in Poland, for instance, the calling-season is the time for this sport, which is so highly esteemed, that in Hungary and Poland a tablet is fixed in the shooting-lodges to commemorate the death of each bird.

THE SQUIRREL

(Sciurus vulgaris)

OF all the smaller animals inhabiting the forests of Great Britain and the Continent, none is better known, more graceful in its actions, or more charming in its appearance than the squirrel, the typical representative of a vast family of rodents, second in number only to the members of the mouse tribe, and having an almost cosmopolitan distribution, although unknown in Australia. The European species, like the great majority of its immediate relatives, is completely arboreal in its habits, and frequents for choice dry and shady forests where there is abundance of tall trees. Its range includes practically the whole of Europe and a considerable portion of northern Asia; but, as might be expected, there is considerable local variation in the matter of colour in different parts of this extensive area, and the British squirrel differs markedly from the typical Swedish representative of the species. There are, moreover, especially in the British race, seasonal changes of colour, which render its appearance very different.

During the winter months the British squirrel (*Sciurus vulgaris leucurus*) is a foxy coloured rodent, with long tufts of hair to the summits of the ears and the tail practically the same colour as the body. In the spring the long winter coat of the head and body is exchanged for a shorter summer dress and the ears lose their tufts; but the hairs of the tail are not changed, and consequently become dirty white, owing to the bleaching effect of light on their colour. This dirty white or cream-coloured tint of the tail in summer is absolutely characteristic of the British squirrel; and it may be added that even in winter this appendage is much less red than in many continental squirrels, being in fact reddish brown.

In the typical squirrel (*S. v. typicus*), of southern Norway and Sweden, the body in summer has a brownish red coat very similar to that of the British race; but the tail is red, and does not bleach when the hairs are old and worn. In winter the body-coat is soft greyish brown in colour, with traces of the summer tint along the middle line of the back and on the limbs. We now come to the subject of our illustration, which shows the German race of the squirrel (*S. v. rutilans*) in its winter coat. In this variety the colour is bright red at all seasons, although there is a tinge of light smoky grey along the flanks in winter; the tail being at all seasons bright rufous, often rather darker than the body. There is, however, a brown phase of this race; and all continental squirrels exhibit a more or less marked tendency towards individual melanism.

Omitting mention of several other red or reddish races of the species met with in various parts of the Continent, reference may be made to the Grecian squirrel (*S. v. lilæus*), in which the tendency to blackness is general if not universal; the general colour being brown, passing into blackish on the hind half of the back and the outer sides of the limbs. This, however, by no means exhausts the colour range of this extremely variable species, for in northern Russia and Siberia we find squirrels (*S. v. argenteus* and *S. v. sibiricus*) in which the general colour of the winter coat is light French grey, with the long ear-tufts black. It is these grey squirrels which are used in such numbers to form the linings of ladies' cheap cloaks; but perfect skins, to say nothing of the living animal, are scarcely ever seen in England.

For the reception of their young, squirrels build a well-constructed nest, or "drey," which is oval in shape, and made of fibres and leaves with a lining of moss; its usual position being the fork of a large tree or a hole in the trunk. The fibres are neatly and intricately interwoven; and when the nest is placed in a fork, the entrance is usually made to open near to one of the branches, with the colour of which it agrees very closely. In this comfortable home the female brings forth three or four young, usually in June, which are tended by both parents, with whom they remain till the following year.

Although squirrels are stated to make an occasional meal of birds' eggs, they are in the main strict vegetarians, feeding chiefly upon pine-cones, nuts, beech-mast, bark, buds, and young shoots. Where they are unusually numerous, as in certain parts of Scotland, they are stated to inflict considerable damage on young larch-plantations. When feeding, squirrels sit up and grasp the food in their fore-paws, with which they hold nuts while these are pierced by the chisel-like front teeth. Long flying leaps from tree to tree are frequently taken by these active rodents.

THE ROE-DEER

(Capreolus caprea)

THE roe-deer, or roebuck, as it is commonly called, is the smallest European representative of the deer family, or *Cervidæ*, and belongs to a small group confined to Europe and northern Asia. So far as external characters are concerned, roe-deer differ from more typical *Cervidæ*, such as the red deer and the fallow deer, by the simpler structure of the relatively small antlers of the bucks, which rise nearly vertically from the head, and carry only three points; the basal, or brow, tines of the red deer being absent. As a distinctive character, common to both sexes, may be mentioned the absence of a tail.

The summer and winter dresses of the species, as in so many of the deer of the temperate zone, are strikingly different; the summer coat being bright foxy red, while in winter the general colour of the fur is olive-brown. At the latter season, at any rate, there is a conspicuous white patch on the rump, which serves as a guide to the hinder members of a family or party when fleeing from danger. The beautiful fawns are marked with a comparatively small number of longitudinal rows of white or yellowish spots and streaks upon a rufous ground; this indicating that the roe is descended from deer of which the adults were similarly spotted in summer. The black moustache-mark on the muzzle, and the white tip to the chin, are other features of these elegant little deer.

The bucks attain their full development in the third year, when the antlers, which commence as simple spikes, first acquire their third tines. Adult bucks usually shed their antlers about Christmas, and the new ones, which increase in size, although not in complexity till the sixth year, are in most cases fully developed by the end of February. The fawns, of which there may be either one, two, or three at a birth, usually make their appearance in the world in May, at any rate in the British Isles.

The favourite haunts of roe-deer are woods and forests on the plains, where under-wood is abundant, from which they issue forth at evening to graze in meadows and corn-lands. On the Continent these deer are, however, also found in forests on the lower mountains, as well as on the spurs of the higher ranges. As a rule, they associate for the greater part of the year in small family parties; such parties, according to continental writers, usually consisting of a buck with two or three does and their fawns, although it has been stated that roe are strictly monogamous. When put up in covert, they generally start off at a gallop with enormous flying leaps, but their speed is not great. They are also excellent swimmers, and can likewise climb rocks to a certain extent. Their food comprises grass, herbs, berries, and the young shoots of bushes and trees; ivy-leaves, where they are to be obtained, forming a favourite article of diet.

The range of the roebuck extends from the British Isles to Spain and Italy in the south and to southern Scandinavia in the north, while eastwards it stretches across Poland and the south of Russia at least as far as the Caucasus. In the Altai and certain other parts of Siberia it is, however, replaced by a much larger, paler-coloured species with more thickly haired ears, commonly known as the Siberian roe (*Capreolus pygargus*), of which a local race inhabits the Tian Shan range. Still farther east, in Manchuria and Mongolia, we come upon a smaller and redder species, the Manchurian roebuck (*C. manchuricus*, or *bedfordi*), which is more like the European roe, the red coat being exchanged for one of olive-brown or grey in winter.

In former days roebuck were doubtless distributed all over Great Britain; but by the middle of the eighteenth century they appear to have been killed off everywhere, except in the highlands of Scotland. Later on, with the increase of game-preservation, they have reasserted themselves, and spread over the lowlands of southern Scotland, as well as parts of the north of England. In North Wales they were reintroduced into Vaynol Park in 1874; and they have likewise been turned down in the Blackmoor Vale of Dorsetshire, where they are now once more wild.

In height, a good roebuck will stand fully 26 inches at the shoulder; while in weight he will turn the scale at 60 lb., if in first-rate condition. On the Continent these deer are generally hunted by means of beating the woods, where the guns are stationed along the tracks by which the roe pass to their feeding-grounds. Roe-venison, which is in best condition during winter, is generally regarded as inferior to that of either the red or the fallow deer.

THE REINDEER

(Rangifer tarandus)

THE reindeer, the ren of the Swedes, is by far the most valuable member of the deer tribe, as it furnishes the Laps and many of the tribes of northern Asia not only with food, raiment, and leather, but likewise serves as a beast of draught and burden to transport them and their food across the inhospitable regions which form their home. Reindeer are likewise in all probability the most numerous in individuals of any of the *Cervidæ*, occurring in vast herds on the high *fjells* of Scandinavia, while in many parts of North America, where they are known as caribou, they are met with in countless thousands, if not indeed in millions.

But it is not only in these two respects that reindeer are worthy of special notice, for they are the only members of the deer tribe in which antlers are carried by both sexes, those of the females being, however, considerably smaller than those of the males; while they are further remarkable for the early period of life at which these appendages make their first appearance. Then, again, the antlers, as is well shown in the illustration, are quite unlike those of any other deer; generally having the two pairs of front tines more or less branched and unsymmetrical, while the main beam sweeps backwards and then forwards in a bold curve, frequently giving off a single back-tine at the middle of the arch, and always carrying a number of tines on the hind edge of the upper portion.

Reindeer have a circumpolar distribution, except that they are naturally absent from Alaska; and in the former respect therefore agree with elk, although their range extends much farther north, and is proportionately curtailed in the south.

In all respects these deer are admirably adapted to a climate of intense severity and a life for months at a time amid snow and ice. Their coats are of great thickness and density, the hairs growing so close together as to produce a structure recalling much elongated velvet-pile. In the stags the throat is further protected by a fringe or ruff of long hair; and in both sexes the main pair of hoofs is very large and deeply cleft, so as to afford as big a surface as possible to prevent sinking deeply in the snow, while further support is afforded by the unusually large size of the small supplemental pair of hoofs. With these powerful hoofs, reindeer in winter scrape away the snow to uncover the reindeer-moss (*Cladonia rangiferina*), which at this season forms their main or only food-supply. In summer, however, they eat grass and herbage, as well as the buds and young shoots of dwarf birch.

The typical representative of the species is the Scandinavian reindeer, the one represented in the illustration, which occurs in both the wild and the domesticated condition. It is a comparatively small animal, with relatively short limbs, and the antlers rounded, and not displaying, as a rule, any very excessive development in the width of one of the lower pair of front tines. In Siberia other and larger, but still imperfectly known, races occur; while in Arctic America the species is represented by two very distinct types, more or less connected by a number of intermediate races.

The recently described Finnish race of the reindeer (*Rangifer tarandus fennicus*), now nearly extinct, is a larger animal than its Swedish representative.

Of the two chief American types, the most northern is the so-called Barren-ground reindeer or caribou (R. *t. arcticus*), in which the antlers are rounded and of great length, with the basal front tines far removed from those of the terminal extremity; while the woodland reindeer (R. *t. caribu*), on the other hand, has the antlers short, flattened, and with the tines crowded together, very large, and often much branched. In some of these American reindeer one of the lower pair of front tines often attains an enormous width, or depth, and is much branched.

There is likewise great racial variation in the matter of colour among American reindeer, the lightest being the Newfoundland R. t. *novæ-terræ*, while the darkest is R. t. *osborni* of the Cascade Mountains, in which the greater part of the body is chocolate, or even blackish brown.

Reindeer, alike in the Old World and in America, are accustomed to undertake long seasonal migrations in search of food, travelling southwards in autumn, and returning to the northern part of their range in summer. Moreover, in many districts they are compelled to retire in summer from the open plains to the shelter of mountain forests on account of the attacks of various insects, more especially the reindeer-fly. Here it may be mentioned that the latter insect is found in certain continental localities far south of the present range of the reindeer, where it has doubtless existed since the time when that animal had a wider distribution. In their migrations, several thousands of reindeer frequently collect in herds of from two hundred to three hundred head, which travel one after another in long lines, each led, it is reported, by an old cow, and the whole forming a very forest of antlers. They swim the widest rivers with ease.

THE DEFASSA WATERBUCK

(Cobus defassa)

TO one of the handsomest of the larger South African antelopes the old Dutch colonists gave the name of wasserbok, the equivalent of the English waterbuck, on account of its partiality for the neighbourhood of water, although it has subsequently been discovered that several more or less nearly allied species are equally aquatic in their habits. This typical waterbuck, which is about the size of an average mule, and is known to naturalists by the name of *Cobus ellipsiprymnus*, is characterised by the iron-grey colour of the long coarse hair of the head and body, and the presence of a large elliptical white ring (whence the specific name *ellipsiprymnus*) on the buttocks. The tail is comparatively long, and terminates in a blackish tuft, while the lower portions of the legs are likewise blackish. The ears are large and rounded; and the bucks carry a handsome pair of sublyrate, light brown horns, ringed from the base nearly to the tip.

At first this was the only waterbuck known, but as the country was gradually opened up a second species was discovered, whose range is now known to extend from Angola and German East Africa to Abyssinia, Senegambia, and Nigeria. This is the defassa or sing-sing waterbuck—for it has different native names in different parts of its range—the *Cobus defassa* of naturalists, and the subject of the accompanying coloured Plate. From the true or typical waterbuck it is distinguished by the redder (or in the case of one race blacker) colour of the coat, and the substitution of a comparatively small white patch for the large elliptical ring on the rump. There are several local races of this species, such as the typical defassa of Abyssinia and the sing-sing of West Africa, differing to a certain extent in colour; the most marked of these being the Angolan *C. defassa penricei*, which is blackish grey.

Waterbuck are very generally found in the neighbourhood of rivers and lakes, where they feed amid the swamps of reeds and papyrus, but are by no means restricted to such situations, and may indeed be met with in dry and rocky localities. Sometimes, like hartebeests, they will climb the white-ant hills in order to obtain a good view of their surroundings and see whether all is safe.

They are gregarious antelopes, usually associating in parties of from three to a score or more; and it is noteworthy that when on the move the troop is invariably led by an old cow, and never by a bull. When danger is declared, the whole herd makes off in a resounding gallop for the nearest water in hope of finding safety by swimming. The flesh of both species of waterbuck is coarse and pervaded by a strong and unpleasant taste, so that it is almost uneatable by Europeans.

In Uganda, where defassa waterbuck are still comparatively numerous, the calves, of which there is usually one at a birth, are born from about the middle of December to the latter part of February. The male calves do not develop their horns till they are about eight months old, by which time the animals are approximately half-grown. At all times shy and difficult to approach, waterbuck are specially wary when they have calves with them. Unlike so many antelopes and nearly all deer, when they take to flight at the approach of danger they scarcely ever turn round, after galloping a certain distance, to gaze at the intruder; and this absence of curiosity saves many of them their lives. The calves, if taken young, are easily tamed.

Waterbuck are the biggest members of a large genus, all the representatives of which are confined to Africa south of the Sahara; the smaller species being known as kobs. A well-known species is Buffon's kob (*Cobus coba*), a nearly uniformly rufous antelope, with blackish fronts to the fore-legs, of the approximate size of a fallow deer, and inhabiting tropical Africa from the west coast to Uganda. Somewhat larger are the puku (*C. vardoni*) and the lechwi (*C. leche*), first discovered by Livingstone in the Zambesi district; both these being foxy-coloured antelopes, without black leg-markings. In the Lake Mweru district of Barotseland there exists another kind of lechwi in which the adult males become blackish brown. And much farther north, in the swamps of the White Nile and the Sobat, we meet with two other members of the genus, the white-eared kob (*C. leucotis*) and Mrs. Gray's kob (*C. maria*), in which the old bucks are likewise nearly black, with the exception of the ears, certain portions of the head and throat, and the under-parts, which are white. These black kobs are highly specialised species; although less specialised than the sable antelope, in which the adults of both sexes are black.

THE HARE

(*Lepus europæus*)

FOR many years the true or brown hare was known scientifically as *Lepus timidus*, but as that name was originally bestowed by Linnæus on the blue hare of Sweden, it has been transferred to that species. Perhaps less confusion would have resulted had it been allowed to continue as the technical designation of the species with which it was so long associated.

No one, of course, can mistake a hare for any other animal, and therefore anything in the way of description would be superfluous. It is, however, important to point out the characters by which the brown, or English species, is distinguished from the blue hare, which is found in Scotland and Ireland.

The brown hare, then, is specially characterised by its relatively large size, its very long ears, which exceed the head in length, the presence of a distinct white streak above each eye, and of a rusty red area on the thigh and generally another on the flank; additional characteristics being the black upper surface of the tail and the black tips to the ears. The general colour of the thick soft fur is rufous tawny, mingled with black above and white beneath, the dark area extending, however, to the throat and chest; there is but little difference between the summer and the winter coat, although the latter is always somewhat the lighter. Length of limb, especially in respect of the hind pair, is a structural characteristic of the hare.

The range of the brown hare includes the whole of central Europe, and a small portion of western Asia; its northern limits in Europe being formed by the lowlands of Scotland, southern Sweden, and southern Russia, while in the south it extends to northern Italy, southern France, and Spain. Nine local races of the species are recognised, of which the British is known as *L. europæus occidentalis*.

Compared with the brown species, the blue or mountain hare, now known as *L. timidus*, is a smaller animal, with the ears, hind-legs, and tail shorter, the head smaller and more rounded, and the colour frequently bluish or brownish grey above in summer, without any rufous or brown on the flanks, but with black tips to the ears, and the under surface of the body white. The upper surface of the tail may be either dark or white; and as a rule the fur, with the exception of the black ear-tips, turns white in winter. The range of this species extends from Ireland and Sweden to the Alps, the Altai Mountains, the Caspian district, and Japan. It is represented by numerous races, of which three occur in the British Isles. Of these, the Irish hare (*L. timidus hibernicus*) has the ears shorter than the head, and the tail wholly white; the coat may turn white in severe winters. In the Scottish hare (*L. t. scoticus*), on the other hand, the ears are nearly as long as the head, the upper side of the tail is dusky in summer, and the whole coat, exclusive of the ear-tips, turns white in winter.

The brown hare is essentially an animal of the open country, and more especially bare fields and fallows, with which its colour harmonises in a wonderful degree. And there is abundant need for this protective resemblance, as the hare has a host of enemies, against whom it has constantly to be on guard, and from whom its sole hope of escape depends upon its limbs. All the three protective senses, hearing, sight, and smell, are highly developed; the long ears detecting every audible sound, while the full, large, round eyes, with widely distended pupils, catch the smallest rays of light at night, when the hare is most active. It has been stated, indeed, that the eyes remain open during sleep, as the eyelids cannot be completely closed; but this is incorrect.

Hares pass most of the day in a lair or "form," which is a smooth place between tussocks of grass or other covert, but they may live out in the open. The females produce at least two litters during the year, the number of leverets in which usually varies from two to five, although it is stated there have been as many as eleven. The young are born quite active and with their eyes open; those which come into the world in spring being capable of breeding the same autumn. The mother remains with her offspring only for the first five or six days after their birth, and then leaves them to shift for themselves.

The young hares of each litter remain together till half-grown, when they disperse; in fifteen months they attain full size, and their average duration of life is seven or eight years. Owing to their long hind-legs hares run much better uphill than downhill.

THE POLAR BEAR

(Ursus maritimus)

LIVING amid eternal ice and snow, the polar bear, which is equalled in bodily size only by some of the huge brown bears of Alaska and Kamchatka, evidently owes its white, or in some instances pale cream-coloured, coat to its surroundings; this white livery, like that of the polar hare, being worn throughout the year. The species is always alluded to simply as the polar bear, although its full title should be the north polar bear; the Antarctic, so far as we know at present, having no land mammals.

In its native haunts the polar bear is found alike on the ice-bound coasts and islands, and on the ice-fields themselves, where it obtains much of its food, this being captured both on land and in the water. Indeed, this great carnivore is fully as much at home in the sea as on *terra firma*, and is capable of swimming long distances at a stretch.

In former days it is probable that the polar bear ranged considerably farther south than is the case at the present day, when it is but rarely seen even in the south of Greenland. The species, like Arctic animals generally, has a nearly circumpolar distribution, and has been divided into a number of local races. These are at present distinguished by skull-characters, but if a sufficient series of skins were available in museums for comparison, there would probably be found local differences in the colour, length, and character of the fur. Polar bear skins are, however, of great commercial value, so that no collection contains a large series of specimens. Moreover, the exact locality of most of the skins offered for sale by furriers is unknown.

Against the intense cold of its Arctic home the white bear is well protected by its long and dense coat, as well as by the thick layer of fat underlying the skin. By means of the hairy covering of the soles of the feet—which in other bears are naked—the animal is enabled to obtain a firm foothold on the ice; upon which, as well as on land, it is a swift and long-winded runner. As special adaptations in the bodily form to swimming and diving, may be mentioned the thin, compressed body, the long neck, the small ears, and the long pointed head, which offer the least possible resistance to progression in water. The strong limbs, with broad paws and webs between the toes, form efficient oars when swimming; while the oily nature of the fur keeps the water off the body. The white bear is, in fact, as admirably adapted to a life among the Arctic ice as is the lion to the deserts of Africa; and both animals may be regarded as the absolute rulers of their respective domains.

In the pursuit of its prey the polar bear displays great craft and ingenuity. When it spies a seal sleeping by a hole in the ice, or on the edge of an ice-floe, if it succeeds in approaching undetected, it glides swiftly and silently into the water, swims a certain distance under the surface, and then rises to observe the situation of its victim. In this manner, by alternate dives and risings, it progresses till within a short distance, when it makes a final dive, to rise near the sleeping seal, which it generally manages to capture. In summer these bears are almost complete vegetarians, subsisting on grass, berries, lichens, moss, and sea-weed. At other seasons their chief food consists of the flesh of seals, walruses, and the smaller cetaceans, such as the white whale and the so-called blackfish. In some districts they capture fish of various kinds, and more especially salmon. In addition to the above, numbers of the smaller polar mammals, such as the Arctic lemming, are caught and eaten; while the young and eggs of various sea-birds, especially auks and guillemots, also form a portion of the diet.

White bears are seldom seen in numbers except where the carcases of whales have been left to rot by the whalers; and generally these animals go about in pairs, accompanied by one or two cubs, which the female will defend with her life.

Only when driven by hunger will the white bear venture to attack human beings. Sealers, who were formerly much afraid of them, nowadays attack the bears armed only with lances, and kill large numbers.

In the far north white bears disappear for the most part during the long Arctic winter, and it is believed that many hibernate, especially as they have occasionally been found in holes. In the winter lair the female gives birth to her tiny, short-haired, and blind cubs, which are usually one or two in number, although triplets occasionally occur.

THE MANDRILL

(Maimon mormon)

THE mandrill is a highly specialised and at the same time extremely hideous West African representative of the dog-faced baboons, nearly all of which are confined to Africa south of the Sahara, although one species is a native of southern Arabia. All these baboons have the long straight muzzles from which the group derives its name, and all except the subject of the accompanying Plate are more or less uniformly coloured animals, with, in most cases, comparatively long tails. The mandrill and its ally and compatriot the drill are, however, distinguished from all their relatives by the reduction of the tail to a mere stump; while old males of the former are further characterised by the presence of large fluted swellings on the sides of the muzzle and the brilliant colouring of these and the other bare parts in this region, while a nearly equal brilliancy is developed in the naked patches on the rump.

To the female mandrill, who, as shown in the Plate, lacks the nasal swellings and brilliant hues of her lord and master, this style of decoration may, and probably does, appear beautiful, but to ourselves it is simply hideous and repulsive, as are the manners and ways of this monstrous ape. The nature of the colouring of the old males is sufficiently indicated in the coloured Plate; but it may be mentioned that the scarlet area on the muzzle has the appearance of vermilion sealing-wax, while the ultramarine of the lateral swellings is suffused in the flutings with shades of violet, more especially when the animal is under the influence of excitement. Indeed, when in this condition, all the colours are intensified and heightened.

In size the male mandrill may be compared to a short-bodied mastiff, while in strength and ferocity it has few equals, so that it is rightly dreaded by all the natives of West Africa. The female is a much smaller and much less powerful animal.

The mandrill, which is one of the shyest of all apes, inhabits the rocky parts of mountain forests in the Gold Coast, Guinea, and the adjacent districts of West Africa. Its food consists of fruits, bulbous plants, grass, and various other herbage, birds' eggs, and all small animals that it can capture.

When captured young, the mandrill is sufficiently amiable, and for a time it remains tame and amenable; but, in the case of males at any rate, in the course of a few years its naturally evil disposition asserts itself, and it soon becomes one of the most vicious and disgusting brutes in creation. Indeed, there is not a good word to be said in its favour. In confinement the rage of the old males is something frightful, and it takes but little to excite them to this frenzied condition, when they shake the bars of their cages, and endeavour to rush upon the objects of their aversion. Little wonder that the West Coast natives dread the mandrill more than they do the lion.

Information is still required as to the habits of the mandrill in a state of nature; and it does not appear to be known whether these apes associate in large droves, after the manner of the ordinary dog-faced baboons, or whether they go about in pairs.

Here it may be mentioned that the name mandrill apparently signifies a man-like baboon, although there is little approximating to the human type in either the physiognomy or the general appearance of this hideous creature; the name drill being an old English word, of which one signification denotes an ape or baboon. By the Germans the mandrill is known as the forest-devil, which is perhaps a more appropriate designation; while by one of the older English naturalists it was termed the rib-faced ape, in allusion to the fluted, melon-shaped swellings on the sides of the muzzle.

The drill (*M. leucophæus*), which is likewise West African, but appears to have a more extensive range in that part of the continent, is a smaller animal than the mandrill, with only small swellings on the face of the old male, which is uniformly black. The bare patches on the rump are, however, bright red; but the tail, which is carried bent forwards over the rump in a similar manner, is hairy on all sides, instead of having its lower surface bare, as in the mandrill. The limbs, moreover, are longer and more slender than in the mandrill; and in fact in all these particulars the drill tends to form in some degree a connecting link between the former and more ordinary baboons.

THE WOLF

(*Canis lupus*)

AS the lion or the tiger forms the supreme development of the feline stock, so the wolf constitutes the culminating branch of the canine line. Each of these animals represents indeed the acme of perfection of which its particular type is capable; but in regard to specialisation the members of the cat tribe stand on a much higher plane than those of the dog family. In the cats we find the face much shortened and rounded, the number of cheek-teeth reduced both at the front and the back of the series, the pair of cutting-teeth in each jaw consisting almost entirely of scissor-like blades, and the claws retractile and protected by large horny sheaths.

In the dog tribe, on the other hand, the long face and numerous cheek-teeth of the more primitive extinct Carnivora are retained, the pair of cutting-teeth are by no means wholly blade, and the claws are non-retractile and devoid of sheaths.

Equally marked differences in the matter of habits are likewise characteristic of the two groups. Cats, as a rule, hunt their prey singly or in couples, stalking it stealthily by the aid of sight, and then making a sudden rush or spring upon the victim. Most members of the *Canidæ*, on the contrary, associate in large packs, which hunt their victims by scent, and pull them down by their combined strength.

Apart from the brown bear, which is of a less active and less carnivorous disposition, the wolf is the largest member of the Carnivora which has been indigenous to central and northern Europe during the historic period; and it plays in the folk-lore of that portion of the continent a part very similar to that taken by the tiger in the legends and mythology of the natives of India. The superstition of the "werwolf" and the familiar story of *Little Red Ridinghood* bear striking testimony to the prominent position held by the wolf in folk-lore.

The wolf, which is in fact the ancestor of many domesticated breeds of dogs, is the largest representative of the *Canidæ*, although the typical wolf of northern Europe is much exceeded in point of size by those of certain other countries, the largest of all being the huge black wolf (*Canis lupus pambasileus*) of Alaska, which is almost as tall as an ordinary bear, and has a head of enormous proportions. Although many modern naturalists divide it into distinct species, the wolf has a circumpolar distribution, and in Asia ranges as far south as the Himalaya and the north-western frontier of India. From Great Britain, Denmark, and Holland the wolf has long since disappeared, and it has likewise been exterminated in northern and central Germany. But wolves still remain in Spain, in the districts of France adjacent to Germany, and in Poland and Russia, as well as in other parts of eastern Europe, while they are abundant throughout central and northern Asia. In the colder parts of their habitat, as in Tibet, where a black phase is common, these fierce Carnivora develop a thick under-coat of fur. With such an extensive range in the Old World, it would be only natural to expect that wolves should present considerable local variation in colour and size; and as a matter of fact this is actually the case, although the various local races are still imperfectly known.

In North America the species is represented by the grey, or timber, wolf (*C. lupus occidentalis*), as well as by the aforesaid black Alaskan race. As regards colour, the ordinary American wolf is dark grey, becoming almost black on the back, with dusky patches on the shoulders and hips; but there is also a more rufous phase. The coyote, or prairie-wolf, is a distinct species, *C. latrans*.

In Europe the favourite haunts of the wolf are wild and retired situations both in the plains and mountains; during spring and summer wolves go about for the most part singly or in pairs, but in winter they collect in parties or large packs, and when driven to desperation by hunger are the terror of the peasantry or the traveller, not only in Russia but even in some districts of France and Germany.

The food of the wolf includes all animals that it can capture and overpower, as well as carrion, and even vegetable substances. Large game and domesticated animals form its chief victims, but in cases of necessity, rats, mice, frogs, and even cockchafers are not disdained. In pursuit of their prey wolves are practically untiring. The female gives birth during the spring, in some sequestered situation in the forest, to from three to nine cubs, which she nurses and tends with wonderful affection and care.

THE WILD CAT

(Felis catus)

SAVAGE as a wild cat is a proverbial expression; and although in attitudes of repose, like the one selected for the coloured Plate, the ferocity of look may temporarily disappear, there are few animals fiercer or more bloodthirsty than *Felis catus*. Indeed, a wild cat brought to bay or caught in a trap is a perfect fiend incarnate.

The wild cat and the lynx are the only feline Carnivora indigenous at the present day to central and northern Europe; and of these two the lynx has long since disappeared from the British Isles, and has become very scarce throughout the greater part of the Continent, but is still common in Scandinavia and Russia. The wild cat, on the other hand, survives in most European countries, and is far from uncommon in the Alps and many parts of France and Germany, although in Great Britain its sole remaining refuges are among the highlands of Scotland, and even there it has been suggested that many of the so-called wild cats are hybrids, or domesticated cats run wild. Eastward, the range of the wild cat extends at least as far as the Altai, where its representative has a larger and more woolly coat than the typical western race.

In Africa the place of the European wild cat is taken by a closely allied but somewhat less strongly striped species known as *Felis ocreata*, of which the northern race was protected and held sacred by the ancient Egyptians in Bubastis and other cities. That the European and the African wild cats have given rise to the domesticated cats of the greater part of the Old World may be considered certain, although there is some difference of opinion as to the respective shares taken by the two wild species in the production of the tame breeds. Among the ordinary house-cats of western Europe, the type coming nearest to the wild ancestor is the one with black transverse stripes on a grey ground. The true tabby, on the other hand, that is to say the type in which the dark markings take the form of broad longitudinal or obliquely longitudinal bands arrayed in a ring-like or spiral manner on the flanks, is unlike either of the two wild cats, and it has been suggested that it may have had a totally distinct ancestry.

By choice the wild cat, which for most of the year lives alone, frequents large, thick, and sequestered pine-forests, where it selects rocky situations as affording the best hiding-places. In addition to clefts in the rocks, it chooses, however, for its lair the burrows of the fox or the badger, or holes in large tree-trunks, while it will sometimes make its abode in thick bushes.

In regard to food, the wild cat preys chiefly upon rats, mice, and small birds, but it also kills larger animals, such as pheasants, hares, rabbits, and squirrels, while it occasionally ventures to attack the fawns of roe-deer and red deer, springing on their backs, and tearing open the arteries of the neck. There is an old monkish Latin line that *"catus amat pisces, sed non vult tingere plantas"* (the cat loves fish, but does not like to wet its feet); but this applies only to the domesticated breed, for when dwelling near streams or lakes the wild cat will capture both fishes and water-fowl. Like many other animals, it will often kill more than it can devour, as if from the very love of slaughter.

The male and female live together only during the pairing-season, and while the kittens are unable to look after themselves. At other times each individual has its own particular haunt, from which, however, it makes long excursions into the surrounding forest or the neighbouring plains; such foraging expeditions often lasting for days together. In winter it frequently deserts the forest to take up its abode in old uninhabited buildings or other safe places of refuge.

In some sheltered situation the female gives birth in spring to five or six kittens, very similar to those of domesticated cats, and likewise born blind.

When captured, wild cats, whether old or young, are impossible to tame; and it is for this reason that the species is so seldom seen in menageries. As already mentioned, it has been suggested that many of the cats found wild in Scotland are hybrids between the wild species and the tame breed, but there is no evidence that such interbreeding takes place; and such of these animals as are not true wild cats seem to be individuals of the domesticated breed which have reverted to the wild state, and have thereby assumed to some extent the characters of the ancestral type.

THE RED KANGAROO

(Macropus rufus)

LITTLE wonder that when Captain Cook and his companions first beheld kangaroos bounding over the plains of Australia they were overwhelmed with astonishment, and regarded them as the most extraordinary animals in the world. They are, indeed, unlike any other creatures, and form the supreme development of the terrestrial and herbivorous section of the marsupial type, being admirably adapted to their environment, and occupying in Australia the place held in other countries by cattle, sheep, antelopes, deer, and hares. For there are kangaroos of all sizes, ranging from the gigantic species, with its 6 foot stature, forming the subject of the present notice, down to the diminutive kangaroo-rats, no larger than rabbits.

That kangaroos, like all marsupials, bring forth their young in an exceedingly imperfect and helpless condition, it is almost unnecessary to mention; as is the fact that the females carry them in the pouch not only till they are able to take care of themselves, but, on occasion, till a much later period. Not that it is to be assumed from this that the young of all marsupials are carried in the maternal pouch; this being in some cases undeveloped, and the offspring in such instances merely clinging for a time to their mothers' nipples.

For a long time it was generally considered that marsupials were survivors of the ancestral stock which gave rise to the ordinary, and so-called placental, mammals; but this is now known to be a mistaken view, and marsupials and placentals are evidently divergent branches from a common ancestral stock. In Australia marsupials have developed almost as many different modifications as have the placentals in other parts of the world; and at first sight there seem few external characters in common to such diverse types as the leaping, bipedal, herbivorous kangaroo, and the cursorial and carnivorous Tasmanian wolf. When, however, the two animals are observed more closely, we may note a curious resemblance in the form of their heads, and, above all, by the great, thick tail, passing almost imperceptibly into the body; both these features proclaiming their comparatively near relationship and descent from a common ancestor.

Among the external characteristics of the kangaroo tribe may be noted the somewhat deer-like head, the short fore-limbs, each armed with five toes, and used in progression only when the animal is grazing, and the enormous length and strength of the hind-legs, in which one toe is greatly developed at the expense of the rest, and serves not only to aid in progression, but forms an offensive weapon of great power and effectiveness. In the sitting posture, as represented in the illustration, the body is supported on a tripod formed by the massive tail and the hind-legs; the whole of the lower portion of the latter, corresponding structurally to the human foot, being then applied to the ground. When, however, the animal is leaping, as shown in the background of the picture, the hind part of the foot is raised and the body supported by the toes; the tail thumping on the ground at each leap.

Kangaroos subsist chiefly on grass and leaves, but also eat buds, bark, roots, and fruits. Originally they were to be found all over the habitable parts of Australia and Tasmania, from the coasts inwards, and while most kinds inhabited the plains, others were to be found in the mountains. Nowadays, however, they have for the most part been driven back into the uncultivated lands of the interior, and their numbers have been greatly reduced. All the larger species associate in large bands. True kangaroos of the genus *Macropus* are also found in New Guinea, while a few of the smaller kinds are natives of the eastern Austro-Malay Islands.

The great red kangaroo, the subject of the illustration, is the largest member of the whole group, and takes its name from the dark rusty red colour of the soft woolly hair of the adult males, that of the females being a delicate bluish grey. Its home is in the rocky districts of South and East Australia.

As this species is kept in most Zoological Gardens, opportunities have been afforded of observing the newly born young. These come into the world after a gestation of only thirty-nine days, when they are only about an inch and a quarter in length, and little more than shapeless lumps of animation. The newborn young is immediately transferred by the mother, by means of her lips, to the pouch, where it is attached by its sucking mouth to a nipple. Here it remains for eight months, after which it returns to the shelter of the pouch, when so disposed, for a considerable time longer: a fresh offspring being often in the pouch at the same time.

THE BLUE ROLLER

(Coracias garrulus)

AMONG the birds of brilliant plumage which occasionally straggle to the British Isles, one of the most gaudy is the blue roller, so called on account of its roling, or "rolling," flight. In India this and an allied species are commonly called the blue jay, while in Poland the present bird is locally termed the Polish parrot. The roller has, however, nothing to do with either jays or parrots, but is more nearly related to the king-fishers.

Although the blue roller is practically an unmistakable bird—certainly so far as British species are concerned—it may be well to mention that the head, neck, wing-coverts, and under-parts are bright greenish blue, the back and shoulders cinnamon-brown, the flight-feathers blue at the roots and elsewhere black above and wholly deep blue beneath, the upper tail-coverts deep ultramarine, while the two middle tail-feathers are dirty brown, and the other feathers of the tail sky-blue, with the tips of the outer pair, which are somewhat lengthened, black. In size the bird is rather smaller than a crow.

The normal summer range of the species includes central and southern Europe, and thence extends through central Asia to Kashmir; while in winter it embraces India and the greater part of Africa. Northwards this bird is found as far as Scandinavia, although only occasionally; but its chief haunts in Europe are Spain, Portugal, Greece, Poland, and southern Russia. It makes its annual appearance in Europe towards the end of April, and takes its departure, like the cuckoo, not later than August. In Asia Minor, Persia, Baluchistan, India, and Ceylon it is replaced, except in winter (when both are found together), by the closely allied *Coracias indicus*. A third species extends from the eastern Himalaya through Burma to Cochin China and Siam, a fourth inhabits Celebes, and quite a number are indigenous to Africa. In Australasia and the New World these birds are unknown.

Avoiding swampy localities and high mountains, the roller resorts to dry open districts with thin forest in which birch abounds, and where hollow oaks or beech are to be found. Here it prefers to dwell at the edge of the forest, where isolated trees grow in the fields and by the roadsides, and command a wide view of the country. In harvest-time it repairs to the sheaves in the cornfields.

The bird is, however, largely insectivorous, and from its perch in some tall tree sallies forth in search of all kinds of insects and other invertebrates as well as young frogs, while it will also eat field-mice. In their season figs afford it a much appreciated feast.

In habits the roller is shy and unsociable, going about in pairs, and generally, but not always, shunning the vicinity of human habitations. Its flight is swift and undulating, and often limited to the passage from one tree-top to another. In fine weather, however, it indulges in the habit of tumbling when on the wing, and performs all kinds of aerial evolutions, sometimes falling from a considerable height almost to the ground; this last performance generally taking place during the pairing-season. At that season, when the males are in the company of the females, the former rise into the air with their characteristic harsh cry of "rak, rak, rak," and descend again with a rapidly repeated "rak, rak, rak." When at rest, the cry is a high-pitched, frequently repeated "raker, raker, raker." It is from these cries that the bird derives its specified name, and likewise its German title of *blaurake*.

In many parts of Europe the nest is built in the hole of a tree, but in the south is more generally placed in ruins, under the roofs of houses, in clefts in walls, or on cliffs and steep banks, in which deep holes are excavated for its reception. At the proper season it contains from five to six white eggs, which are brooded in turn by the male and female, and this so assiduously that it is often easy to capture the sitting bird with the hands. In laying white eggs, the roller conforms to the general rule of birds which nest in holes. It is, moreover, an uncleanly bird, and when the nest has contained young for some time, it gives out a most disgusting smell, while the young themselves sit in a mass of filth. Strange that such a beautiful bird should have such dirty habits!

As regards its mental powers, the roller is evidently highly developed; but its inherent shyness renders it unsuitable as a cage-bird. In disposition it is, as already mentioned, unsociable, and it will frequently quarrel with other birds and its own fellows. On the other hand, several pairs often nest near together, and its migrations are made in large companies.

THE BITTERN

(*Botaurus stellaris*)

BEFORE drainage and cultivation had driven away so many of the marsh-haunting birds, the boom of the bittern was a familiar sound to the dwellers in the fens of Cambridgeshire and Lincolnshire, and this handsome bird regularly nested not only in those districts, but in the Norfolk Broads, as well as in many other counties possessing situations suitable to its habits. Indeed nests were taken now and then up to the middle of last century, and even later; but at the present day the bittern is nothing more than a casual visitor to the British Islands, in fact, so rare have its visits become, that they are generally considered worthy of special record.

A near relative of the heron, the bittern does not expose itself in the open after the fashion of that species, but skulks amid the shelter of reeds and flags, where its presence is made known only by the ordinary raven-like croak, or the loud booming of the male in the breeding-season. From this habit the bittern derives its German name of *rohrdommel*; and with such surroundings its mottled plumage of light and dark brown mingled with black is designed to harmonise, as is also in all probability the greenish hue of its long, spear-like beak, so admirably adapted to seize and hold the unwary fish or frog that may come within striking distance.

As this resemblance between the plumage of the bittern and its inanimate surroundings is sufficiently apparent from the accompanying Plate, it will be unnecessary to attempt any description of its colouring. It is, however, important to mention that the bird appears to be in the habit of increasing the protective power of its mottled livery, by assuming, probably under the influence of alarm, a statuesque position amid the reeds, with the body held as erect as possible, the neck stretched to its fullest extent, and the head and beak pointing skywards. In this posture it is stated, by those who have had the good fortune to see it, to be almost invisible amid the upright brown stems of reeds and bulrushes.

The range of the bittern is very extensive, including the whole of temperate Europe, northern Africa, and the greater portion of Asia lying between the Himalaya and the Arctic Circle, as well as north-western India and Burma. In South Africa its place is taken by another species, a third kind inhabits Australia, New Caledonia, and New Zealand, while a fourth is a native of North and Central America, and a fifth is indigenous to tropical South America. The distribution of the group is thus almost cosmopolitan, if we except most of the tropical zone of the Old World; and all the five species are closely related.

In Europe at the present day bitterns are still common in Spain, Holland, many parts of France and Germany, and the swamps of the Danube and Volga. It should be added that in Africa, India, and Burma the species occurs only in winter, being to some extent migratory in its habits.

These shy birds associate in pairs, but in migration time collect in parties previous to setting out on their travels. The nest, which is built of reeds and flags, and lined with grass, is well concealed among the brake in which the pair have taken up their residence, and may be a floating structure. The eggs, which are from three to five in number and greenish blue in colour, are incubated solely by the hen, who is supplied with food by her mate. The chicks remain in the nest until fully fledged, and display extraordinary cleverness in making their way among the reeds.

Little comes amiss to the bittern in the way of food, so long as it is of an animal nature; and its appetite is so great, that it is on the hunt for prey from sunrise till sunset. Water-rats, field-mice, birds of all kinds, fish, especially those inhabiting muddy water, snakes, lizards, leeches, worms, and insects and their larvæ are alike devoured by these voracious birds.

The usual cry of the bittern is, as already mentioned, a hoarse, raven-like croak; and it is only during the breeding-season that the male utters the resounding boom, which has been compared to the bellowing of a bull, and on a still night may be heard at a distance of a mile or more. How this wonderful sound is produced is not yet definitely known.

A wounded bittern is most dangerous to approach, as it will strike with unerring aim at the eye of the gunner who approaches to seize it; and the formidable beak is likewise employed, and generally with success, to repel the attack of any fox bold enough to approach its owner.

It should be added that during the hot summer of 1911 a pair of bitterns bred in the old Norfolk haunts of the species.

THE KAFIR CROWNED CRANE

(Balearica chrysopelargus)

THE crowned cranes of Africa, of which there are three species, constituting by themselves an exclusively African genus, are some of the handsomest members of a beautiful and stately group of birds; the fan-like array of bristly feathers on the head, which constitutes the so-called "crown," conferring on these cranes a regal appearance which is lacking in their smooth-headed relatives.

Of the three species, one (*Balearica pavonina*) is a native of north-eastern Africa, ranging southwards into West and Equatorial Africa; the second, forming the subject of the accompanying Plate and sometimes known as *B. regulorum* instead of by the designation here used, is a southern bird, ranging from the Cape to the Zambesi and Loanza valleys; while the third (*B. gibbericeps*) hails from East Africa.

As regards the distinctive features of these three cranes, the northern species is greenish black above and dark grey below, with most of the feathers of the sharply pointed lanceolate type; the neck is delicate pearl-grey; the secondary quills are chestnut, and the wing-coverts partly white and yellow; the twisted bristly feathers forming the crown are yellow and white with black tips; a bare area on each side of the face is white above and pink below; and black down clothes the throat. This species has a very small wattle on the throat; but in the Kafir crane this wattle, as shown in the Plate, is much larger and mainly red in colour. The southern species is further characterised by the greyer tone of the plumage of the upper-parts, and the white cheek-patch, with only a margin of crimson above. In the East African *B. gibbericeps* this cheek-patch becomes much larger than in either of the other two species, extending backwards nearly to the nape of the neck.

The carriage of these handsome, well-built birds is upright; while their gait, when they are walking with measured steps, is calm and stately. These birds can, however, run with great speed, so fast indeed that a man can keep up with them only with difficulty. The flight is heavy and slow, with powerful, measured strokes of the wings, and the neck and legs stretched out. The crest, or crown, is at the same time depressed. When in flight, a party of cranes always arrange themselves in wedge-shaped form in order to cleave their passage through the air with the least possible resistance; a very powerful bird taking up the position at the apex of the triangle. As in all cranes, the cry is very loud and resounding; its piercing, trumpet-like notes being due to the complicated structure of the windpipe, which is arranged in coils, and consists of more than three hundred bony rings.

The food of these cranes consists chiefly of various kinds of grain and other seeds, among which those of durrha or Kafir millet form the largest portion. The seed-spikes of grass, buds of trees, and fruits, as well as a certain proportion of insects, are, however, also devoured by these birds.

Very curious are the attitudes assumed by these handsome birds, especially when they are surprised or under the influence of excitement. On such occasions they place themselves in wonderful postures, bending their bodies up and down, spreading out their wings, and then joining their fellows in a dance, during which they often spring a yard high in the air, while all the time their feet are alternately lifted and put down in regular rhythm.

The daily life of the crowned crane displays great uniformity. At sunrise the whole flock flies out into the veldt, where its members search for food at least a couple of hours; then they proceed to the water to drink, and spend the day in digesting their morning's meal. Their favourite resort for the day is a sandbank far out in a river, where they can stand and preen their feathers or doze without fear of interruption. At evening the flock returns to its sleeping-quarters in the forest.

All the graceful habits of these cranes can be studied in Europe, where the birds flourish in confinement, if given sufficient room. In captivity they soon learn to associate with the human beings and animals with whom they are brought into contact, and are specially keen in discriminating between those who treat them well and those who dislike their companionship. These cranes are also to be met with in every Kafir village, while they are likewise frequently tamed by the European settlers in both South and East Africa.

In Europe the group is represented by the ordinary grey crane (*Grus cinerea*) and the elegant demoiselle crane (*G. virgo*), both of which lack the crests of their crowned relatives.

It will be observed that in this notice the name "Kafir" is spelt with one *f*; this being the proper orthography, as the word is the Arabic "Kafir," an unbeliever, this being exemplified in the name "Kafiristan," the land of infidels.

THE SILVER GULL

(Larus argentatus)

IN its adult plumage, with the snow-white head, neck, and under-parts, the delicate French grey back and wings, and the white-spotted black tips to the larger flight-feathers, the silver gull, or herring-gull, as it is more commonly called, is one of the most beautiful members of a lovely tribe of birds. Indeed, whether swimming calmly on the surface of the sea, or skimming over the crests of the waves borne on their long and powerful pinions, and every now and then plunging into the water to seize a fish or some floating morsel of food, gulls in general are some of the most elegant and graceful of all birds, their delicate colouring, in which grey and white, relieved to a greater or less extent by black or chocolate, generally predominate, thus giving a refinement to their whole appearance which is wanting in many birds of brilliant plumage. Were it not that their cries, their tempers, and their habits are by no means angelic, gulls might well have been selected as emblems of the angels.

The white and pale grey plumage, replaced in a few species by a wholly white or cream-coloured livery, is, however, developed only in the adults; birds of the year having the back and wings thickly mottled with brown and dark grey, and the tail black, while the head and neck are wholly brown; the beak, moreover, in the species forming the subject of the Plate, being black instead of orange. From this we learn that gulls are descended from birds with relatively dark plumage, which may perhaps have been dwellers on the land; and if this be so, these beautiful birds evidently acquired their present type of colouring only when they took to a life on the ocean wave. In any case, the pale livery of the adult gull must be regarded as a special adaptation to its mode of life; such a garb being the one which harmonises best with the foam-flecked waves of the waste of waters.

The herring-gull closely resembles in colouring the common gull (*Larus canus*), and like that species is abundant on the British coasts, or, for that matter, on the Thames at London Bridge or the ornamental water in St. James's Park in winter. It is, however, a much larger bird, attaining a length of about 22 inches in the case of adult males.

This species, moreover, is much less intolerant of heat than the common gull; and while the former is compelled to wing its way to the more northern coasts for the breeding-season, the herring-gull, like the kittiwake, nests by scores on the southern coast of England, wherever conditions suitable to its habits exist. The kittiwake, it may be mentioned, differs from other gulls by the absence of the hind-toe, and is therefore referred to a genus by itself, under the appropriate name of *Rissa tridactyla*. The only other species likely to be confounded in the summer plumage, when the black-headed species have donned their chocolate or black caps, is the greater white-winged gull (*L. hyperboreus*), which differs by the paler tone of the plumage generally and more especially by the feature to which it owes its name.

The herring-gull is a wide-ranging species, met with on both sides of the North Atlantic, extending eastwards to the White Sea, and in winter as far south as the Black and Caspian Seas and the Mediterranean. In America herring-gulls visit in summer the inhospitable coasts of Labrador and Greenland, but in winter wander south to the genial climate of the West Indies and Central America, where, in all probability, they cross the continent to join a closely allied gull inhabiting the Pacific. In Europe the southern breeding range of this handsome species seems to be formed by the northern coasts of France.

Herring-gulls, where conditions are favourable, may be found nesting on the coasts of the British Isles from the south of England to the Orkneys and the Shetlands, as well as in Ireland, where they are the most common and most widely spread members of their tribe in the breeding-season. Sometimes only a few gulls nest in company, but in other situations large colonies collect for breeding purposes; and it is noteworthy that the breeding sites are always in the neighbourhood of the shore and generally on tall cliffs.

Another noteworthy feature of the herring-gull is that the adult livery is not assumed till the fourth year, in consequence of which an unusually large number of birds in the speckled dress are always in evidence.

THE GREAT HORNED OWL

(Bubo ignavus)

THE great horned, or eagle, owl, the largest European representative of the nocturnal birds-of-prey, is the typical member of a group characterised by the relatively small size of the apertures of the ears, which are not closed by covers, and likewise by the more or less imperfect development of the disc-like ring of feathers round the eye, which forms such a conspicuous feature in ordinary owls. The so-called horns are, it need scarcely be mentioned, tufts of long, somewhat hair-like feathers growing from the neighbourhood of the ears. These ear-tufts are common to the so-called long-eared owls, which are, however, all birds of smaller size, with larger ear-openings, protected by special covers, and complete discs of feathers round the eyes.

The general appearance of this magnificent bird when in an attitude of repose, the colour and markings of the plumage of the head, breast, and under-parts, and the great, staring, red-ringed eyes are admirably shown in the illustration. When, however, the bird is enraged, the body is depressed and the plumage ruffled out, while the wings are half-spread, thus increasing its size and producing a formidable appearance probably sufficient to overawe a number of would-be assailants other than man. In length this unmistakable bird measures as much as 27 inches, so that it is equal in point of size to a small eagle.

The eagle-owl ranges over the greater portion of Europe, as well as northern Africa, and much, if not the whole, of northern and central Asia; and the species breeds as far north as Lapland and as far south as Gibraltar and Greece. It is true that the eagle-owls inhabiting the country to the east of the Ural Mountains have been regarded as a distinct species, under the name of *B. sibiricus*, while the name of *B. turcomanus* has been proposed for those from the deserts of south-western Siberia, Turkestan, and Tibet; but these and others from Asia north of the Himalaya are so closely allied to the European bird that they are best regarded in the light of local races of that species.

On the other hand, the American eagle-owl (*B. virginianus*) is a perfectly distinct species, with a range extending over the whole of North America, although this bird has likewise been split up into a number of nominal species. Eagle-owls of various species are also known from tropical South America, the whole of Africa, Arabia, India, China, and Japan, so that with the exception of the Malay countries and Australasia, the group has a practically world-wide range.

To the British Isles the great horned owl—the *grand duc* of the French and the *uhu* of the Germans—is, nowadays at any rate, merely an occasional straggler, and then only to the northern parts of the kingdom, most, if not all of the specimens that have from time to time been taken in England being birds that have escaped from captivity. There is, however, a report that these splendid owls once inhabited the Orkneys.

Eagle-owls thrive well and breed freely in captivity; years ago a number were kept, for instance, at Arundel Castle, but these, although long regarded as European birds, were eventually proved to belong to the North American species. Captive specimens have served to demonstrate in some degree the great age to which these owls will live; a female brought from Norway in 1827 having survived for seventy-five years in an English aviary, and having produced during the last thirty years of her captivity no less than ninety offspring.

From its large size, powerful beak and claws, and fierce disposition, the eagle-owl, which is mainly nocturnal in its habits, is a terrible foe to the smaller animals, preying largely upon rats, mice, moles, birds, and frogs, while it also attacks and devours larger game, such as roe-deer fawns, hares, rabbits, hazel-hens, ducks, and geese. Its haunts are thick forests, especially those in which tints of rufous grey and black prevail; and in such situations the mottled, warm-coloured tints of its plumage harmonise both with the bark of the tree-trunks and the weathered surface of the rocks and cliffs. At even, with noiseless wings, the eagle-owl issues from its perch to sweep over the plains in search of prey; or, rising high in the air, utters its loud screech, and awakes the slumbering birds, which flutter from their roosts to meet their fate by the relentless talons and beak of the nocturnal marauder.

The huge nest is generally built in trees, on cliffs, or in old buildings, but the two or three eggs may be laid in holes on the bare ground. The eggs are brooded by the female, who is fed during incubation by her partner.

THE FLAMINGO

(Phœnicopterus roseus)

FLAMINGOES, of which there are several species, present us with one of the most striking instances of adaptation to a particular mode of life to be met with in the animal kingdom, more especially as the main feature in this adaptation is developed in its full perfection only when the bird is mature and takes to a special diet. In common with other waders, the adult flamingo has an enormously long neck and legs, and is thereby enabled to procure its food from depths inaccessible to most other birds, although it is frequently content to search for food in the shallows. Its distinctive structural peculiarity is, however, the sharp downward flexure of the extremity of the beak, and more especially that of the lower half. Such a beak appears at the first glance quite unsuited for groping up food from the mud of marshes and lagoons which form the favourite haunts of these stately birds; but while thus engaged, flamingoes turn their heads the wrong way up, when the beak at once becomes a most efficient ladle, admirably adapted for collecting and holding the small spiral univalve molluscs of the genus *Cerithium* which in many districts form their chief food.

In the young flamingo, whose diet is of a different nature, the beak is more or less normal in form.

Flamingoes, with certain relationships to the storks, appear to have most affinity with ducks, geese, and swans; and it is curious to note how like are the head and neck of a flamingo, if the beak were but straightened out, to those of a swan, the resemblance extending in some cases even to the colour of the beak,—red or orange at the base and black at the tip.

White and scarlet, or crimson—the Easter colours—are the colours of the flamingo, but the relative proportions of these vary according to the species. The European flamingo—the subject of the Plate—whose range extends from central Europe to the Canaries and the Cape Verd Islands, and thence all over Africa, and eastwards to Lake Baikal, India, and Ceylon, has, for instance, the greater part of the plumage white or pinkish white with scarlet wing-coverts and black quills, red legs, and the beak pink at the base and black at the tip. Much more gorgeous is the tropical American *P. ruber*, ranging as far south as Para and the Galapagos Islands, in which the general colour of the plumage is light vermilion, with brighter wing-coverts, the base of beak being yellow and the legs red. To the south of central Peru, in Uruguay, and perhaps in Brazil this species is replaced by *P. chilensis*, distinguished by the legs being grey with red joints, while the black of the beak extends upwards above the bend. In all the foregoing species a hind-toe is present, but this is lacking in two other South American species, namely, *P. (Phœnicoparrus) andinus*, the largest member of the family, of the Chilian and Bolivian Andes and Argentina, and *P. (P.) jamesi*, of southern Peru and Chile, in both of which the beak is yellow at the base and red in the middle, while the legs are yellow in the former and red in the latter. Lastly, there is *P. (Phœniconaias) minor*, of Africa, Madagascar, and India, which in general appearance much resembles the European species.

Although flamingoes spend much of their time in wading, they are also good swimmers. Like geese and ducks, they associate in vast flocks, and further evidence of their kinship to that group is afforded by their loss of the power of flight during the height of the moulting season, and likewise by their "gaggling" cries, which are curiously like those of geese.

In the breeding-season flamingoes resort to lakes, salt-lagoons, or the swamps in river-valleys, those of the Guadalquivir being one of their favourite haunts in Europe. Bare shores are an essential element in such breeding-colonies, for it is on these that the birds construct their curious sugar-loaf mud-nests, which have a cup-like depression at the summit for the reception of the one or two eggs, and vary in height from two to fifteen inches according to the depth of the water. The eggs, which are brooded by each sex in turn for a period of fully four weeks, have bluish coloured shells, covered with a rough chalky crust. When incubating, the bird sits with its legs bent beneath the body, although it was long supposed that these hung down on the sides of the nest.

Hundreds or even thousands of flamingoes may congregate in these breeding-colonies; and there are few more beautiful sights in nature than to see a flock of these splendid birds, especially the scarlet American species, rise on the wing and display their full colouring and plumage in the sunlight.

THE NILE CROCODILE

(Crocodilus niloticus)

TO the ancient Egyptians the timsa, or Nile crocodile, the champsa of Herodotus, was a familiar reptile; but from the lower and middle portion of the Nile, as far up as Thebes, the species has long since been exterminated, although it is still abundant in East and South Africa. Elsewhere it occurs in Madagascar, and still survives, although sparingly, in Syria, more especially in the Zerka (= Crocodile) River near Cæsarea. In Biblical times crocodiles were, however, abundant in the Holy Land; and there is little doubt that the "leviathan" of the Book of Job refers to these noisome reptiles. Till nearly the fifteenth century it also seems that crocodiles lived in Greece and the Isles of the Grecian Archipelago; and at that date a huge crocodile's skull was jealously preserved and exhibited at Rhodes. It is these ancient south European crocodiles which probably gave rise to the legend of St. George and the Dragon and other myths of a kindred nature.

In popular estimation there is much confusion between crocodiles and alligators; and in India the former are almost invariably called alligators, although there is not a single representative of that group in the whole country. Indeed, alligators are confined to China and America; those of South America, properly known as caimans, differing from the typical Mississippi species by having a bony armour on the under as well as on the upper surface. There are many characters distinguishing crocodiles from alligators; one of the most easily recognised being that the fourth lower tooth of a crocodile bites into a notch on the outer side of the upper jaw, so that its summit is visible when the mouth is closed, whereas in an alligator the corresponding tooth is received into a pit, so that the tip is completely concealed when the reptile shuts its enormous mouth.

True crocodiles, which, with alligators, caimans, and gavials, are the largest of living reptiles, are now represented by about eleven species, whose combined range includes Africa, southern Asia, northern Australia, and tropical America. India possesses two species, the broad-nosed muggar (*Crocodilus palustris*), and the narrow-nosed estuarine crocodile (*C. porosus*); the former of which ranges to Ceylon, Burma, and the Malay Peninsula and Islands, while the latter, which enters salt-water freely, and is sometimes found far out to sea, extends from India, Ceylon, and the south of China to northern Australia and the Solomon and Fiji Islands.

Crocodiles may grow to a length of 18 and probably 20 feet or more, and are powerful creatures, which are the pests of the waters they frequent, seizing and dragging down women and men who come to draw water, and gripping by the nose cattle and other animals which come to drink. A beast thus seized has little chance of escape, no matter what may be its size, as it is taken at a disadvantage and rendered comparatively helpless; but there is a well-authenticated story of a Nile crocodile seizing a rhinoceros by one of the hind-legs just as it was about to leave the river, and eventually dragging the enormous beast, despite its frantic efforts, backwards into deep water, where it was drowned.

The prey is, indeed, always killed by drowning, after which the carcase is dragged ashore and concealed among reeds or other covert, where it is left until decomposition has set in before being devoured. A large proportion of the food of the Nile crocodile is stated, however, to consist of fish; and snails, water-fowl, and carrion are also devoured by these ravenous reptiles. When in repose, crocodiles lie like logs in the water, or on the neighbouring sand and mud banks, but the slightest sound awakens them to activity.

Hearing seems indeed to be the sense most strongly developed in crocodiles; and it is upon this that they depend in ascertaining the presence of prey. Smell, touch, and taste appear to be but poorly developed; and the tongue is affixed to the lower surface of the mouth throughout its entire length. Crocodiles are furnished with glands secreting a musky substance, much esteemed by the Sudanis for anointing their hair and bodies, and thus producing what is to them, no doubt, an agreeable odour.

All crocodiles lay hard-shelled, oval eggs, about the size of those of a goose. These are deposited by the female, to the number of from about twenty to ninety, a yard or so deep in the sand of the river-bank, where they are carefully covered up by means of her tail, and then left to mature by the action of the sun's rays. The mother appears, however, to keep watch in the neighbourhood of the nest, and when the eggs are ready to hatch, their occupants utter a peculiar sound which attracts the parent to their assistance.

THE WHITE-HANDED GIBBON

(Hylobates lar)

THE name gibbon, which is of unknown origin, and appears to have been first used by the French naturalist Buffon, is applied to the members of the smallest representatives of the man-like apes, all of which differ from their larger cousins, the chimpanzi, gorilla, and orang-utan, by retaining small bare patches on the buttocks. In this respect they show evidence of kinship to ordinary monkeys and baboons, in which, however, these naked callosities, as they are called by naturalists, attain much greater development.

Gibbons agree with the larger man-like apes in the complete absence of the tail, but are distinguished by the great length of the limbs, more especially the front pair, so that when these curious pigmies are standing upright on their feet, their hands touch the ground. It is very interesting to note in this connection that gibbons are the only man-like apes which habitually assume the erect posture in walking, when the arms are usually elevated and in some cases extended horizontally, thus giving a most remarkable appearance to their owners.

All the members of the group are confined to tropical south-eastern Asia, where they range from the eastern portion of the Himalaya through Burma and the Malay countries to the island of Hainan, off the southern coast of China. The largest species, the siaman (*Hylobates syndactylus*) of Borneo, does not much exceed one yard in height. In colour they are very generally black, often, as in the species represented in the Plate, with a narrow white band or fillet surrounding the greater part of the face, or making merely a line above the eyebrows. In the white-handed species, forming the subject of the illustration, the white fillet attains its fullest development, and the hands and feet are likewise white. This species inhabits the mountain-forests of Tenasserim at elevations between three thousand and five thousand feet, whence it extends through the Malay Peninsula. Its near relative the hulok (*H. hulok*), whose range extends from Assam to Arakan and the neighbouring districts, lacks the white hands and feet, while the fillet on the face merely forms a band over the eyebrow. In the Hainan gibbon (*H. hainanus*) even the eyebrow-band is wanting.

A specimen of this last-named species, which is very near akin to the hulok, afforded a surprise to naturalists in that it changed the colour of its coat during its sojourn in the London Zoological Gardens. When obtained in 1897, at an age of about six months, its colour was dark smoky grey, but this soon changed to black, which persisted till after the animal's arrival in London in January 1904. In February of that year the fur began to assume a greyish tinge; and by autumn the entire coat had become silvery, or stone, grey, except for a black stripe down the middle line of the head. It does not, however, appear that this colour-change is universal in the species; for another captive specimen believed to be about a dozen years old was wholly black.

This is confirmed by the observations of a French traveller in the case of the white-cheeked gibbon (*H. leucogenys*) of Annam and Siam, in which two distinct colour-phases are noticeable.

These gibbons commonly go about in parties of six or seven; and while some individuals are black with white whiskers and a band on the sides and lower part of the face, in others the general colour is golden yellow with no white on the face. That the two phases are specifically identical is certain, and it does not appear that the light-coloured individuals eventually become dark, or *vice versa*. The natives, it is true, have an idea that the light-coloured individuals are the females, but this seems to be disproved by the fact that they are much less numerous than the black ones; and it thus appears probable that the species exhibits two distinct colour-phases.

Gibbons are almost wholly arboreal apes; and in the trees they move about mainly with the aid of their long arms, by means of which they swing themselves from bough to bough, and thus from one tree to another, so that a whole troop will traverse the forest without descending to the ground. So active and agile indeed are these apes, that when confined in open cages they will catch birds on the wing, which apparently form part of their food. In addition to these, gibbons subsist on fruits of various kinds, leaves, young shoots, insects, and spiders.

All gibbons have a remarkably powerful double-note cry, like *hu-lok*, which has a wailing sound, and awakes the echoes in the early morning and again at evening in the forests inhabited by these apes. At a distance the cry has some faint resemblance to the human voice; but at close quarters it is enough to drive the traveller, when resting in his tent, well-nigh distracted.

THE ABYSSINIAN GREEN MONKEY

(Cercopithecus æthiops)

AS a familiar example of a very large assemblage of exclusively African long-tailed monkeys collectively known as guenons (from a French word signifying to make grimaces), no better species could have been selected for illustration than the Abyssinian green monkey. This monkey, which like its cousin the common green monkey (*Cercopithecus callitrichus*, or *sabæus*) may be seen in almost all menageries and on many street-organs, is an East African species; whereas the true green monkey is West African, ranging from Senegambia through Sierra Leone to northern Liberia, but has been introduced into some of the Cape Verd and West Indian Islands, where it still occurs plentifully, at least in Barbados.

From the guereza monkeys (*Colobus*) of Africa, a guenon may be distinguished at a glance by its well-developed thumbs; while these monkeys also differ from guerezas, as well as from their Asiatic relatives the langurs, by having well-developed pouches in their cheeks for storing food. There is, however, another African group of monkeys, the mangabeys (*Cercocebus*), some of which are more difficult to distinguish from guenons, although the majority are recognisable at a glance by their flesh-coloured eyelids, whence the name of white eyelid monkeys. When this character fails, resource must be had to the cheek-teeth, of which the last pair in the lower jaw is less complex in a guenon than in a mangabey; but to examine the back-teeth of a live monkey is not an experiment every one would desire to try!

To any one but a specialist, the very large number of representatives of the guenons are exceedingly difficult to class and identify. The matter has, however, been somewhat simplified by arranging the species in thirteen groups, each characterised by the possession of some particular distinctive feature or features. The well-known diana monkey (*C. diana*) represents, for instance, a group in which there are long, upwardly directed, snow-white whiskers associated with blackish or dark red under-parts, a white brow-band, distinctly white areas on the chest and the inside of the arms, a chin-tuft, and a white stripe on the outer side of the thigh. The members of the spot-nosed group, again, as typified by *C. petaurista*, are recognisable at a glance by having a large, heart-shaped white spot on the tip of the nose. The green monkeys, on the other hand, belong to a group (typified by the Abyssinian species) somewhat less easy to recognise, but characterised by the absence of black on the outer surface of the arms, which may be coloured like the body, but are usually, as in the plate, somewhat paler and grey. The face is nearly always black, but may be mottled or freckled with pigment; and, except in one species, the whiskers grow upwards and backwards over the ears. The western species, the aforesaid *C. callitrichus*, lacks the white browband present in the Abyssinian species and nearly all the other members of the group; the species forming the subject of the plate being further distinguished by the black-speckled yellowish olive colour of the fur of the back, the long, white whiskers, sharply defined from the hair of the rest of the head, and the presence of a tuft of hair at the root of the tail.

Guenons may be regarded as some of the most typical of all monkeys, and those in which monkey-tricks attain their fullest development. They live more on fruits and seeds than either the guerezas or the langurs, which feed largely on leaves and the young shoots of trees. Guenons are indeed probably to some extent omnivorous, whereas the members of the other two groups are wholly herbivorous. This difference in the matter of diet is correlated with the presence of cheek-pouches and the simple character of the stomach in the guenons, whereas the other two groups lack pouches in the cheek, but, in compensation, have the stomach folded and divided in a complex manner.

In their native forests green monkeys and their relatives associate in large troops, which keep up a constant chattering, and each of which is under the leadership and control of an old male. Each troop appears to have its own particular territory in the forest; and if one party intrude on the domain of its neighbour, a fierce contest takes place, which does not end until the invaders have been driven out or have proved themselves the stronger. A survival of this custom may be noticed even among monkeys in a menagerie, where each species or individual will take up one portion of the cage for its own particular use, and resolutely defend it against the other occupants of the enclosure.

From the other members of the genus the patas monkey (*C. patas*) and its relatives are specially distinguished by their red colouring and large size, and they should perhaps form a genus by themselves.

THE FOX

(Canis vulpes or *Vulpes alopex)*

THE fox enjoys a well-earned reputation for cunning and shrewdness, and especially for its capacity to adapt itself to changed conditions and its presence of mind and resourcefulness in emergencies. Nevertheless, it is possible to deceive even a fox. A well-known American naturalist relates, for instance, how he attracted a fox to within a few yards of his own position by imitating the squeaking of a field-mouse. For some distance he came cautiously skulking between the big tussocks of coarse grass, but for the last fifty yards he had to traverse open sward. When he had got half this distance he became suspicious, and began to think that the motionless, squatting naturalist was not a stump or a rock, whereupon, it is curious to note, that he at once ran to leeward of the suspicious object, thus showing that foxes, like many other animals, trust more to their sense of smell than to their sight.

In spite of the fact that everybody knows a fox by sight, there is a difference of opinion among naturalists as to whether the fox of North America is specifically identical with the European animal. The differences between them are, however, so small and insignificant that the most sensible course is to regard the fox as a circumpolar species, with several local races, both in the Old World and in North America. To attempt anything in the way of description of such a well-known animal would be altogether superfluous, although it may be well to mention that the tip of the tail is always white or whitish; and that, as in most members of the dog family, there is a dark-coloured patch on the upper surface of the root of the tail, indicating the position of a gland.

Although nearly related to wolves and jackals, foxes are distinguished by their slighter make, the relatively shorter limbs and longer body and tail, as well as by the sharper muzzle, and also by a peculiarity in the shape of the projecting process of bone which forms part of the upper border of the socket of the eye. In wolves, jackals, and dogs this process is convex, whereas in all foxes it is concave; thus conclusively proving that the fox has nothing to do with the parentage of the dog.

The typical fox of northern and central Europe is one of the largest representatives of the species, and is further characterised by the deep "foxy" red colour of its coat, and the light colour of the under-parts. In southern Europe this race is replaced by one of rather smaller dimensions (*Canis vulpes melanogaster*) and less brilliant red in colour, with the under-parts dark, and sometimes a cross-shaped dark mark on the shoulders. Nearly allied is the Himalayan fox (*C. v. montanus*); but in central Asia and some of the neighbouring countries we find a larger and paler race known as *C. v. flavescens*. No representative of the species is found to the south of the Himalaya or in Burma and the Malay countries; but the central Asian, or a nearly allied, race probably extends right across Siberia and Manchuria.

In North America the fox reappears, and ranges as far south as Georgia; the common and most widely distributed race being known as *C. v. fulvus*, while separate names have been given to the races respectively inhabiting Nova Scotia and Newfoundland. The so-called "cross fox" of America is merely one with a blackish stripe down the back and another across the shoulders, as in many Himalayan and Afghan foxes; while the "silver fox" is a grey, and the "black fox" a black phase of the ordinary American race.

Skins of black foxes are of enormous value, really fine specimens selling for as much as £200 each; they are imitated by dyeing the skins of the white

Arctic fox (*C. lagopus*) and leaving the end of the tail and the tips of the longer hairs white. Although most of these black skins come from America, black foxes apparently also occur in Siberia.

No matter how far north its range, the fox never turns white in winter. Another curious fact about this animal is that the skin, when viewed under the microscope, exhibits a structure indicating that the ancestors of the species were apparently clothed with scales instead of hair; thus affording important testimony to the view that mammals are the direct descendants of reptiles.

Unlike jackals and wolves, the fox is a more or less solitary animal, living for the greater part of the year alone, or in company with the vixen, as the female is called; the latter name being merely the west country pronunciation of *fixen*, which may be the Saxon plural of *fix*. Chiefly nocturnal in their habits, foxes may take up their residence either in holes, or "earths," dug by themselves, in ravines or clefts among rocks, or amid coarse grass and bushes. The young, usually from five to seven in number, have slaty grey coats quite different from their parents. Although foxes are sad enemies to the poultry-yard and the game-covert, it should be remembered that they are death on field-mice and rats.

THE BROWN BEAR

(Ursus arctus)

WITH the exception of the hyænas and the civets, all the more important families of land Carnivora are represented in northern Europe; the wild cat and the lynx doing duty for the *Felidæ*, the wolf and the fox for the *Canidæ*, the pine-marten, glutton, and otter, together with several smaller species, for the *Mustelidæ*, and the bear, or brown bear, as it is commonly called, in order to distinguish it from its relatives—white, grey, or black—for the *Ursidæ*. It is further noteworthy that, with the exception of the wild cat, all these animals are either common to the northern portions of the two hemispheres or are represented in North America by species very closely allied to their European prototypes.

The bear is one of the undoubtedly circumpolar species, for although distinct names have been given to its American representatives, with the exception of the grisly bear (*Ursus horribilis*), which is specifically different, all these are obviously nothing more than local forms or races of a single variable and widely distributed species.

Even in the Old World there are many distinct races of the brown bear. In Syria, for instance, we have a greyish race (*U. arctus syriacus*); while in Kashmir the coat is very generally deep cream-colour, although it tends to darken in old individuals of this race, which is known as *U. a. isabellinus*. From the British Isles the bear has long since been exterminated, but it survives in the wilder parts of Spain, France, Germany, and Hungary, while it is still comparatively common in many parts of Scandinavia and Russia, whence it extends right across Asia north of the southern flank of the Himalaya to Kamchatka and Japan; its southern limit to the east of the Himalaya being marked by the Siamese race (*U. a. shanorum*). Central Asian brown bears, of which the local races are not yet properly determined, generally have light-coloured coats, with a white gorget, which is wanting in Scandinavian and Russian bears. The largest Old World race is the Kamchadale brown bear (*U. a. piscator* or *lasiotis*), of which skins may measure as much as 9 feet in length.

On the opposite side of Bering Strait Alaska vies with Kamchatka in claiming the record in point of size among brown bears, while in the matter of local variation the representatives of the species put the rest of the world into the shade, for American naturalists recognise no fewer than half a dozen different races in Alaska alone. These include the huge Kodiak bear (*U. a. middendorffi*) of Kodiak Island, the Alaskan Peninsula bear (*U. a. gyas*), the gigantic Yakutat bear (*U. a. dalli*) from the neighbourhood of Yakutat Bay and the coast for an undetermined distance north and south, the Sitka bear

(*U. a. sitkensis*) of Sitka and Baranoff Islands, the Admiralty bear (*U. a. eulophus*) of Admiralty Island, and Kidder's bear (*U. a. kidderi*) of the Alaskan Peninsula. All these are, however, so similar to one another, that it requires an expert to distinguish them.

The grisly bear (*U. horribilis*) is, however, a distinct species, characterised, among other features, by its white claws. Typically a native of the high forests of the Rocky Mountains, the grisly is represented by two Alaskan races, namely, *U. h. phæonyx* of the interior, and *U. h. kenaiensis* of the Kenai Peninsula and adjacent coasts. It perhaps reappears in the Tian Shan range, in the heart of central Asia, in the shape of the so-called *U. leuconyx*.

Bears are the least carnivorous of the Carnivora, subsisting in many districts almost entirely on fruits, nuts, shoots of trees, roots, honey, and ants, with an occasional meal of carrion. In Alaska, however, when the bears wake up from their winter sleep and come abroad they find the rivers almost choked with Pacific salmon which are ascending to spawn, and in these they find an abundant and nourishing food-supply upon which they rapidly grow fat. The duration of the winter-sleep, which may take place in a hollow tree, a cavern, or a cleft in the rocks, depends upon the length of the winter; and it is when they first issue forth from these lairs that bears are in their best coats, the summer fur being short and comparatively valueless, while the skins are also difficult to preserve owing to the fatness of the animals at this season. The females bring forth their cubs, one or two in number, during the period of hibernation.

Bears are to a great extent diurnal mammals, whose haunts are chiefly the forests, although in late summer they resort in many localities, such as Kashmir, to the open mountain pastures, where they may be seen grazing at no great distance from flocks of sheep or goats. The adults go about singly or in pairs, the females generally accompanied by their cubs, which may include those of two successive seasons.

THE PINE-MARTEN

(Mustela martes)

THE marten, or pine-marten, as it is commonly termed in order to distinguish it from its continental relative the beech-marten (*Mustela foina*), is one of those species which are on the verge of disappearing from England and Wales. In the midland and south-eastern counties of England the species appears to have become scarce during the first half of the last century, and in most of these it was killed before 1860, although isolated instances of its occurrence in Hertfordshire, Surrey, and Sussex have been recorded since that date. In Lincolnshire, Norfolk, and Suffolk it survived, however, until the 'eighties; and there have been recent occurrences in Leicestershire. The Lake District and the western side of north and central Wales are at present the strongholds of the species; and the isolated occurrences in other parts of the country appear to be largely due to its wandering habits.

In the wilder districts of Scotland, wherever there is sufficient forest, martens are still far from uncommon; and they are comparatively abundant in parts of Scandinavia, France, Spain, Germany, Hungary, and Russia. Among the characteristics of the species, the most easily recognisable is the yellow or orange area on the throat; this part being white in the beech-marten. That these two animals are distinct species, and not merely local races of a single one, seems to be demonstrated by their occurrence in the same districts. How far eastwards the range of the marten extends does not appear to be ascertained. In Siberia the species is, however, represented by the closely allied sable (*M. zibellina*), which in turn gives place in North America to the American marten (*M. americana*). All three might perhaps, however, be regarded as geographical races of one and the same species. Martens are essentially forest-animals, which prefer evergreen trees, and are especially partial to woods where dead tree-trunks lie rotting on the ground or form natural bridges from one stem to another. Here they are as active as squirrels, and chase one another in the same sportive manner. Much of their prey is, however, taken on the ground, and they are deadly foes to hares, rabbits, pheasants, and partridges, as well as to small birds of all kinds. They are also reported to eat beech-mast, and if this be the case, berries may also very probably form a portion of their diet.

The persistency with which a marten will follow the trail of a hare or a rabbit, even in deep snow, is a character the species possesses in common with the weasel tribe generally; and if it fail to pounce upon the unsuspecting victim in the first few springs, it will settle down to the chase as steadily as a beagle or a harrier.

Martens are devoid of the unpleasant odour of the polecat, or foumart, and are therefore suitable for taming. In their native forests they associate in pairs, and make their nests high up in tall trees, selecting in some instances a hole in the stem, but in other cases taking possession of the deserted nests of wood-pigeons or other large birds, or occasionally the drey of a squirrel. In this nest are born early in the spring the six or seven young, which come into the world blind, and are tended by their parents with assiduous care; and they remain in its neighbourhood till they are nearly full grown.

In North America it has been observed that once in about every eight or ten years martens more or less completely disappear from districts in which they are at other times abundant. Strange to say, there is no evidence of the prevalence of disease on such occasions; and it is certain that the martens do not migrate after the manner of lemmings. It has, however, been noticed by the trappers that just before their disappearance none of the martens in the district will take a bait, although at other times they are caught with comparative ease.

In the forest martens and sables are much less difficult to approach than many other wild animals; but they nevertheless display a keen aversion to the neighbourhood of human dwellings, which they never voluntarily approach. Martens and sables, unlike their smaller relative the stoat or ermine, do not turn white in winter; for the very sufficient reason that, on account of being arboreal animals, they would be extremely conspicuous in such a livery. They are so relentlessly hunted for the sake of their valuable fur, that it is a marvel they have not long since been exterminated from many portions of their habitat. The value of the skins of the ordinary European marten varies from eight to twelve shillings, but Russian and Siberian sable commands a much higher price in the market.

THE LEOPARD

(Felis pardus)

IT is not a little remarkable that three such closely related animals as the lion, the tiger, and the leopard or panther (for these two names really indicate the same species) should exhibit such striking dissimilarity in their type of colouring. The explanation of the difference in this respect is, however, probably to be found in the diversity of habit. Neither the lion nor the tiger can climb; and as the former is mainly an inhabitant of open and more or less desert country, it has assumed a uniformly sandy coat to accord with the tawny hue of its surroundings. The tiger, on the other hand, resorts more to covert, and has therefore adopted a livery suitable to such situations, or, at all events, one which tends to break up the general outline of the body.

From both its larger cousins the leopard differs by its arboreal habits, and this may to a great extent account for its spotted type of colouring, which harmonises well with the chequered shade thrown by the foliage on the horizontal and gnarled branches of forest-trees. It is, however, important to notice that the dark markings on the fur of an ordinary leopard are not solid black circular spots, but take the form of rosettes of irregular black patches enclosing a centre of darker tint than the general ground-colour of the fur. Such a type of colouring seems to be one specially adapted to a partially arboreal life, for in the purely terrestrial hunting-leopard (*Cynælurus jubatus*), as well as in the smaller African cat known as the serval (*Felis serval*), the markings all take the form of solid circular black spots. It is further noteworthy that the Amurland race of the leopard (*F. pardus villosa*), which inhabits a country where the forests are coniferous and therefore ill-adapted for climbing by a large animal, the spots are in the main solid and more or less nearly circular.

Leopards have a range larger than that of either the lion or the tiger. In former times they probably inhabited a considerable portion of Europe; and at the present day they extend from the Caucasus over the greater part of Asia, inclusive of Persia, India, China, and the Malay countries, although unknown in northern Siberia, the highlands of Tibet, and Japan. In Africa they inhabit the whole continent from Morocco to Cape Colony, although from certain districts they have now been killed off. As a rule, the central chestnut area forming the centre of the rosettes does not carry any small spots, but two or three such spots occur in the rosettes of some Asiatic leopards; and skins thus marked serve to connect the leopard with its near relative the jaguar (*F. onca*) of Central and South America, in which these central spots are constant. Now the jaguar is an even more arboreal animal than the leopard; and the more complex development of its rosettes is a further proof that this type of marking is one specially adapted to a life among thick-foliaged trees. It may be added that the near relationship existing between the leopard and the jaguar affords conclusive evidence that large spotted felines at one time inhabited North America.

With such an enormous geographical range, it would be only natural to expect that the leopard should present considerable local variation; and, as a matter of fact, such is actually the case. The Amurland, or Manchurian, race has already been mentioned. Asiatic leopards have large rosettes; and among these the Persian leopard (*F. pardus panthera*, or *tulliana*) constitutes a race characterised by its pale colour, long fur, and thick tail. This race makes an approach to the snow-leopard (*F. uncia*) of the highlands of central Asia, which is, however, a distinct species, distinguished by its long fur and tail, with the prevailing colour grey, and the rosettes in the form of ill-defined rings; this type of colouring being probably an adaptation to a life among rocks and snow.

Leopards from North and East Africa show large rosettes of the Asiatic type; but in most other parts of the African continent the rosettes show a great tendency to break up into small spots. Among these small-spotted leopards the Somali race (*F. pardus nanopardus*) is noticeable for its pale colour and small size; its total length being less than 5 feet, whereas some of the other races may measure 7½ feet, or even more, in length.

Both in Asia and north-eastern and East Africa black leopards are not uncommon; but these are merely a dark (melanistic) phase of the ordinary leopard. Nevertheless, it appears, so far as can be judged from a single authenticated instance, that black leopards when mated together will breed true, although, if paired with spotted individuals, the progeny may be of either type.

With regard to the general habits of leopards, space permits only of reference to their remarkable partiality for the flesh of dogs, which is so strong as to render it difficult to keep these animals in leopard-infested districts.

THE LYNX

(Felis (Lynx) lynx)

WHETHER the lynx is more keen of vision than other animals is more than problematical; but, nevertheless, it takes its name from the Greek word (*lukē*) for light, and lynx-eyed has become a proverbial expression. Moreover, as if to set a seal of endorsement upon the proverb, the oldest and most renowned scientific society in Italy bears the title of *Regia Lynceorum Academia* (the "Royal Academy of Lynxes").

Be all this as it may, the lynx is an aberrant and well-marked member of the *Felidæ*, distinguished, together with its immediate relatives, from other members of the cat tribe by its short, stumpy tail, the long tufts of hair at the tips of the ears, and the bushy whitish whiskers fringing the face, as well as by the large size of the feet, and a generally stout appearance, due to the thickness of the coat.

The beautiful fur of the lynx, which commands a high price in the market, has also a character all its own, being long, soft, and silky, with a greyish or reddish tawny colour, and in most cases a number of more or less dark markings, which take the form of longitudinal streaks along the back, of flecks on the sides, and roundish spots on the limbs; the greater portion of the absurd apology for a tail, together with the backs and tufts of the ears, and a conspicuous streak on each side of the face, being black. Many lynx-skins are, however, more or less completely devoid of the dark markings; and it does not yet appear to be ascertained whether the difference in this respect is individual or seasonal.

When the markings are fully developed, they so completely break up the outline of the body as to render the animal almost invisible at a short distance; this being noticeable even within the cramped confines of a cage.

The lynx is the second and largest species of the cat tribe inhabiting northern Europe, where it is common in Scandinavia and Russia, while to the south it is found in mountainous districts as far as Spain. Eastwards it extends across Asia, from the Caucasus and northern Persia, to the northward of the Himalaya to the island of Saghalin. Its fossil remains prove that it was formerly a native of the British Isles.

The Tibetan lynx, on account of its generally paler colour, has been separated as *F. lynx isabellina*, while the one from the Urals, Caucasus, and northern Persia is distinguished as *F. l. cervaria*. The Altai lynx, characterised by its unusually long coat, the, at least frequent, absence of spots, except on the limbs, and its extremely pale colouring, approximating in many parts to white, has likewise received a separate racial name, *F. l. wardi*.

In North America the so-called Canada lynx, although often regarded as a distinct species, differs, in the opinion of a well-known American naturalist, from the lynx of northern Asia and Europe merely by features which may with safety be ascribed to local environment, and should therefore be reckoned as yet another geographical race, under the name of *F. l. canadensis*. Its range extended in former days from Arctic America to the mountains of Pennsylvania; the lynxes of Alaska and Newfoundland respectively forming two minor races.

Lynxes are extremely savage animals with immensely powerful limbs, which appear out of proportion to the lean body. They are inhabitants of forest, where they pass much of their time reposing on the horizontal boughs of trees, whence they can launch themselves with unerring spring on any unsuspecting animal that may be passing below, or from which they may ascend higher in search of squirrels or birds and their nests. They by no means, however, confine their attention to such arboreal game, but are relentless pursuers of hares and still smaller terrestrial mammals of all kinds, while their agility is such that a half-tamed individual has been repeatedly seen to catch pigeons out of a flock feeding on the ground as they rose on the wing. In pursuit of ground-game they advance in long silent leaps, and in winter are prevented from sinking in the snow when progressing in this manner by the great size of their paws.

Tangled thickets or other kinds of dense undergrowth are selected by the lynx for its summer lair; and in such localities in Norway the fluffy cubs are born. In Tibet, however, these animals are compelled to exchange clefts in rocks for the forests of Scandinavia and Russia. Unlike bears, lynxes, in common with all members of the cat tribe, remain active throughout the winter, being sufficiently protected against cold by the thickness of their soft fur. At this time of year they are, however, frequently reduced to great straits by hunger; and their endurance must be simply marvellous to enable them when in this famishing condition to hunt and capture from time to time such prey as may be on the move.

THE INDIAN RHINOCEROS

(Rhinoceros unicornis)

ALTHOUGH formerly ranging a considerable way down the peninsula, the great Indian rhinoceros, as this species should properly be called, is more or less completely restricted to Nepal and the country east of the Tista valley, especially the plain of Assam and Kuch-Behar. Here the mighty beast, which stands over 5½ feet in height at the shoulder, dwells in the tall grass-jungles, where it is as completely concealed as is a rabbit in a meadow ready for mowing. The rhinoceroses, in fact, make for themselves in this giant grass, tunnels, or "runs," in which they move from place to place perfectly secure from observation, and likewise protected from the direct rays of the sun; and it appears that, except to drink, they seldom leave this wonderful covert. To attempt to shoot such enormous beasts on foot in jungle of this description, where escape from the beaten track is well-nigh impossible, would be little short of madness; and the Indian rhinoceros is therefore always hunted on elephants.

In old books on Indian sports the rhinoceros is depicted as charging the elephants, and attempting to spear them with its horn, if not in the act of goring their bodies. This is, however, erroneous, as none of the three named species of Asiatic rhinoceros use their horns in this manner, but employ for offence their sharp, triangular lower tusks, with which they make lateral thrusts and lunges after the manner of a wild boar. African rhinoceroses, on the other hand, have no tusks, and consequently have to rely on their horns—always two in number—for both attack and defence.

It must not, however, be assumed from this that the absence of tusks is compensated by the development of two horns, for there is one Asiatic species, commonly known as the Sumatran rhinoceros (*Rhinoceros sumatrensis*), which has two horns combined with lower tusks; this species thus being the most formidably armed member of the whole group. This so-called Sumatran rhinoceros also occurs in some of the eastern districts of India, as does likewise the third Asiatic species, commonly known as the Javan rhinoceros (*R. sondaicus*), which resembles the great Indian species in carrying but one horn.

A satisfactory and easily recognised distinction between the rhinoceroses of south-eastern Asia and the two African species is afforded by the circumstance that while in the former the hide is thrown into a number of deep folds dividing it into separate areas, in the latter these folds are more or less lacking, so that the skin is comparatively smooth like that of a pig. The shape of the skin-folds serves to distinguish the species forming the subject of the plate from the other one-horned Asiatic species. Among the distinctive features of the present animal may be noted the coif-like expansions of skin at the sides of the head, the large tubercles, recalling the heads of the rivets in an iron boiler, on the shoulders and hind-quarters, and the somewhat triangular shape of the great shield on each shoulder, the fold forming the upper border of which does not extend across the back.

By the older naturalists, rhinoceroses, elephants, and hippopotamuses were grouped together under the title of pachyderms,—a name which has completely dropped out of use in natural history. And very rightly, since the three groups of animals brigaded together under that designation have but little in common with one another. Elephants, for instance, form a group by themselves; hippopotamuses are cousins of the pigs, and thus related to deer and cattle; while rhinoceroses, together with tapirs and horses, form a third group by themselves.

There may seem to the man in the street little in common between a great lumbering brute like a rhinoceros and a Derby winner; but the difference is due solely to the one being a modern specialised type cut out solely for speed, and the other an old-fashioned creature suited for wallowing in marshes or wandering on open plains where it is sufficiently protected by its size and ferocity. Take away the two side-toes from each foot of a rhinoceros, lengthen its limbs, lighten its head and body, modify to a comparatively slight degree its cheek-teeth, and replace its bare "pachydermatous" covering by a thinner, hairy skin, and we should have a horse. Fortunate it is for the naturalist that such primitive creatures as rhinoceroses and tapirs have survived to the present day to afford us an adequate idea of what their numerous extinct relatives looked like in life.

Rhinoceroses are purely herbivorous animals, but whereas the great Indian species subsists chiefly on bamboo-leaves and other grasses, its two Asiatic relatives depend more upon boughs and roots; this difference being correlated with the structure of their teeth. Producing but one offspring at a time, and that at long intervals, these animals apparently live to a very great age, although by no means so long as elephants. The idea that the hide of the Indian species is bullet proof is altogether erroneous.

THE BISON

(Bos [Bison] bonasus)

ALTHOUGH the name bison indisputably belongs to the single species of wild cattle now surviving in Europe, it came to be applied almost exclusively to its American cousin, while the bison itself, till comparatively few years ago, was generally known in this country as the aurochs. The aurochs was, however, the ancient wild ox of Europe, the ancestor of many of our domesticated cattle; and on its extermination, which took place in Poland during the seventeenth century, its name became transferred to the bison, or zubr as it is called in Russia and Poland. Of late years, however, matters have been put right in this respect, and the bison has once more come into its own.

The European bison, although lacking the enormous mass of long hair on the head and fore-quarters which gives to the bulls of its American cousin such a magnificent appearance, is a far better built animal than the latter, as the hind-quarters do not fall away in the same manner. The bulls are considerably larger than the cows, and it is in that sex alone that the beard and throat-fringe, as well as the long hair on the fore-quarters, attain their full development. In summer a short and sleek coat is donned; the old winter coat falling off in spring in large, blanket-like masses.

The bison, apart from a certain number which have been introduced into private parks, now survives only in the forest of Bielowitza, in Lithuanian Poland, now the Russian province of Grodno, and in the Caucasus. In Bielowitza the bison exist in a partially protected condition, and are regularly fed during winter; their number diminished from nearly two thousand in 1857 to three hundred and seventy-five in 1892, and a few years previously had been just over a hundred less. At the present time the herd, which forms an imperial preserve, appears to be suffering from the effects of inbreeding, so that an abnormally large number of the calves are males.

On the other hand, the bison of the Caucasus, which has been separated as a distinct race under the name *Bos bonasus caucasicus*, exists in a purely wild state, although very little information with regard to its numbers, distribution, and habits is available. In former times bison, as we know from historical evidence, as well as from rude although often spirited sketches, believed to be the work of prehistoric man, on the walls of caverns, as well as from their actual fossil remains, were distributed over the greater part of Europe, including Spain.

Here it should be mentioned that the European and the American species are the only living kinds of wild cattle properly entitled to be included under the name of bison; the so-called Indian and Burmese bison belonging to a totally different group of the ox tribe.

European bison, which associate in large herds, are essentially forest animals, although the forest they inhabit must be comparatively thin and intersected with glades and meadows where the herd can graze. The neighbourhood of water is also essential; and when muddy pools are available, they are utilised for wallowing. In addition to taking these mud-baths, bull bison in summer are fond of rolling in sandy or dusty spots.

Leaves and grass form the greater portion of the food of the Bielowitza herd in summer, one particular kind of grass, known as zubr-grass, being a special favourite and communicating to the animals themselves an aromatic odour. Young shoots, twigs, and bark, especially those of the ash, are also largely eaten; and in obtaining bark the stems of trees are frequently stripped as high as the animals can reach, while numbers of saplings are trampled down. Except where they are artificially fed, bison have to depend almost entirely upon buds, twigs, bark, and patches of dry grass and bracken for sustenance; and at this season the herds leave the damper parts of the forest to take up their quarters in the driest situations they can find.

Although an adult bull bison is more than a match for any wolf that ever existed, the Bielowitza herd is constantly harassed by the attacks of wolves, bears, and lynxes, which kill many of the calves, and probably also overpower weakly and half-starved cows in winter, especially if the wolves hunt in packs.

Bison feed chiefly in the early mornings and evening, although they may be seen abroad at all hours. During the breeding-season, in August and September, fierce combats take place between rival bulls for the mastery of the herds. When disturbed or alarmed, the herd breaks into a quick trot, which soon develops into a heavy lumbering gallop; and there is scarcely a finer sight in the world than a charging troop of these magnificent beasts.

THE GAZELLE

(Gazella dorcas)

ALTHOUGH many kinds of gazelles are now known, the gazelle *par excellence*, that is to say, the ghazal of the Arabs, is the beautiful little species represented in the Plate, which, in order to distinguish it from its relatives, naturalists have designated the dorcas. And no more beautiful and delicately made creature exists in the world than this same gazelle, which has formed the theme of poets—especially in the East—for centuries, as the emblem of beauty, elegance, and fleetness. Many people persist in confusing gazelles with deer, although the two groups have but little in common, being broadly distinguished by the characters of their horns, which in gazelles are hollow, unbranched sheaths of true horn supported on cores of bone, while in deer they are branching structures of bare bone. Indeed, the so-called horns of deer are not really horns at all, at all events from the point of view of the naturalist, but rather antlers; and gazelles constitute a section of that group of ruminants collectively known as antelopes.

Of the approximate size of a roebuck, the true, or dorcas, gazelle, like most of its kindred is coloured to harmonise with the more or less desert conditions of its home. The delicate rufous fawn of the upper-parts accords with the yellow tint of the rocks or sand amid which these beautiful creatures spend most of their time; while, when the animal is standing in the full glare of an Eastern sun, the white of the under surface counteracts the effect of the dark shade thrown by the body, and thus, even at comparatively short distances, makes for more or less complete invisibility. Neither is the white "blaze" on the rump without its special use, as it serves as a guide to the members of a troop to follow the line taken by their leader when safety depends solely upon fleetness of foot; the effect of the danger-signal on such occasions being increased by the elevation of the tail, of which the white under surface is then shown. In the bucks alone do the gracefully curved and heavily ringed black horns attain their full development; those of the does being thin and nearly smooth spikes.

The range of the gazelle is large, including the whole of northern Africa, from Morocco in the west to Egypt in the east, and extending southwards to Nigeria and the Egyptian Sudan; while in Asia it embraces Palestine and Syria. In western and southern Arabia it is, however, replaced by the Arabian gazelle (*Gazella arabica*), which is itself a relative of the edmi gazelle (*G. cuvieri*), of Morocco, Algeria, and western Tunisia; the last-named being a mountain-dweller, whereas the dorcas is largely a native of the plains.

In the deserts and plains between the Nile and the Red Sea, where the dorcas gazelle is especially numerous, its food chiefly consists of mimosa-bushes; these bushes growing most abundantly on low, boulder-strewn hills which form the favourite feeding-places of the gazelles. Unlike many ruminants, gazelles are almost constantly on the move, resting only during the very hottest hours of the day, when they seek the shade of the mimosas: and when thus reposing, even experienced hunters find it extremely difficult to distinguish them from the boulders amid which they lie.

Gazelles mostly associate in herds of variable size; but they are generally seen on their feeding-grounds either in small parties of from two to eight head, or singly. They are extremely shy and watchful, and nearly always feed with the wind behind them; while their favourite stations are on elevated ground, where they can command an extensive view of the plains below. When a herd is alarmed and takes to flight, its members always seek shelter on the nearest hill.

The senses of sight, smell, and hearing are all highly developed in gazelles; and the speed of these beautiful antelopes is little short of marvellous. Indeed, when a gazelle is fleeing from the slughi hounds, or so-called Persian greyhounds, with which the Arabs hunt the fawns, they seem scarcely to touch the ground, and to be flying rather than running.

Previous to the pairing-season the bucks fight among themselves, and so fiercely that they not infrequently lose a horn. After a gestation of between five and six months, the doe gives birth to a single fawn, which is at first extremely helpless, when it is most assiduously nursed by the mother. Young gazelles are beautiful little creatures, which in their own country can be kept and tamed without difficulty; in cold climates they require, however, shelter and protection in cold and bad weather.

Pictures of the gazelle are common among the ancient Egyptian frescoes, especially in the temples at Giza, Thebes, Sakhara, and Beni-hassan; and from a painting in a tomb at Sakhara it may be inferred that herds of these graceful ruminants were kept in a half-tamed condition in Pharaonic times.

THE MOUFLON

(Ovis musimon)

THE mouflon, or muflon, is the only wild sheep inhabiting Europe, where it is restricted at the present day to the mountains of Corsica and Sardinia. In former times it doubtless had a more extensive range, and there are reports of its occurrence within the historical period in Greece and the Balearic Islands. It is likewise reported to have once inhabited the mountain ranges of central Spain.

Like all wild sheep, the mouflon has a hairy coat, while its tail is short and deer-like, as in the small domesticated sheep of Soa and other Hebridean islands, of which it is probably the ancestor. The woolly fleece and long tail of many domesticated breeds must accordingly be regarded as features due to careful selection; and in this connection it is important to notice that some of the domesticated sheep of Africa, as well of the East, possess hairy coats like their wild ancestors.

The general colour of the mouflon, like that of so many ruminants, is of a protective nature, being dark above and white beneath, with a white rump-patch as a signal-mark for the members of the flock when in flight. In accordance with the nature of its surroundings, the mouflon has a darker and more rufous coat than the gazelle; but old rams have a whitish saddle-patch on the back in the winter coat. In their native haunts these sheep are stated to be very difficult to detect. The white streaks on the face so characteristic of the gazelles are wanting, but the dark flank-band dividing the fawn of the back from the white of the belly is a feature common to mouflon and many species of gazelles.

In height the mouflon ram stands only about 27 inches, so that the species is one of the smallest of the wild sheep. The horns of the rams usually curve forwards on the sides of the face; the right horn thus forming a right-handed spiral, and *vice versa*. In some Sardinian rams, however, there is an alteration in the direction of the upper part of the spiral, so that the horns curve backwards over the neck, instead of forwards by the sides of the face. The ewes of the Sardinian mouflon appear to be generally, if not invariably, hornless; but some Corsican females, at any rate, carry small horns; and it may be that the presence or absence of horns in this sex forms a distinction between the races respectively inhabiting the two islands.

Mouflon are found only in certain parts of the mountains of Sardinia and Corsica; and when in repose usually resort to high peaks or ridges whence a wide view can be obtained. Moreover, they frequently select situations where currents of air from two different situations combine, and they are then absolutely unapproachable by the sportsman. In much of the ground they frequent, the valleys are filled with a thick growth of ilex; while they feed on the hills amid abundant heather, which affords admirable covert for the approach of the stalker.

In the pairing-season, which takes place during December and January, the old rams, like the males of most ruminants, engage in fierce combats for the possession of the females. The ewes give birth to their one or two lambs during April and May; and these, like the lambs of domesticated sheep, are able to run with their mothers at a very early age. When the rams are in good condition, mouflon-mutton, if hung long enough, is excellent for the table.

In Cyprus, Asia Minor, and Persia the mouflon is replaced by the red sheep (*Ovis orientalis*), a redder and, in the case of some races, larger species, often with a fringe of blackish hair on the throat of the old males, and the horns in that sex always curving backwards behind the neck. The race of this species inhabiting the Troödos mountains of Cyprus is smaller than the rest.

Farther east, namely, in the Kopet Dagh range dividing Persia from Turkestan, the red sheep gives place to the well-known urial (*O. vignei*), in which there is a long white throat-fringe to the old rams, whose horns curve forwards by the sides of the face. This species ranges through Baluchistan and Afghanistan to the Salt Range of the Punjab, and thence along the ranges flanking the Indus into Ladak and Tibet.

In the Altai and Tibet, together with other parts of central Asia, we reach the country of the great argali sheep (*O. ammon*), with its numerous races; while in the Yana Valley of Siberia and in Kamchatka we first meet with the so-called bighorns (*O. canadensis*), of which the typical race is North American. One peculiar species (*Ammotragus lervia*) inhabits the north of Africa, but in the rest of that continent, as also in peninsular India, the Malay countries, and South America, wild sheep are unknown.

THE RED DEER

(Cervus elaphus)

THE red deer, the typical representative of the family *Cervidæ*, is the largest and handsomest European member of that group, although it attains its maximum development in point of bodily size and massiveness of antlers only in eastern Europe and south-western Asia. As in most members of the deer tribe inhabiting temperate countries, there is considerable seasonal difference in the colour of the coat, and the fawns differ remarkably in this respect from their parents. There are also distinctions in colour between the various local races of the species. The ordinary name refers to the fact that in summer a more or less distinct rufous colour prevails on the upper-parts. Here it may be remarked that deer do not in most cases present that marked contrast between the upper and the lower surfaces of the body so characteristic of gazelles and many other members of the antelope group. And the reason for this is not difficult to explain. As mentioned in the text accompanying the plate of that species, the white under-parts of the gazelle are for the purpose of counteracting the dark shadow thrown by the body when standing in full sunlight, and thus to render the animal inconspicuous. Deer, on the other hand, are in the main nocturnal and forest-dwelling creatures, and this type of protective colouring would therefore be useless in their case. The chital, or Indian spotted deer, is, however, much less nocturnal than most species, and also feeds to a great extent in the open; and it is interesting to notice that, in accordance with such habits, this species is white-bellied.

Notable features in the red deer are the shortness of the tail, and the straw-coloured patch on the buttocks in which that brief appendage is included; the same features recurring in its near relative the wapiti. As in all similar cases among ruminants, the light rump-patch serves as a guide to the members of a herd; the place of this being taken in certain other species, such as the fallow deer and the American white-tailed deer, by the pure white under surface of the tail, which is raised when the animals are running.

The antlers of the stag are characterised by the number and regular arrangement of the tines; and more especially, in their fullest development, by the duplication of the first, or brow, tine, and the cup-like arrangement of the terminal snags.

Two features indicate that the red deer is what naturalists term a highly specialised animal. These are, firstly, the shortness of the tail, and, secondly, the white-spotted coat of the fawn, so utterly different from that of the adults of the typical western representative of the species. In the race inhabiting the Caucasus yellowish spots are, however, frequently observable in the coats of full-grown hinds, while similar spots may be developed in adult stags of the North African race,—the so-called Barbary deer. These features clearly indicate that the red deer is descended from a species which was fully spotted at all ages.

The range of the red deer includes, with the exception of the far north, practically the whole of Europe, as well as Asia Minor and part of Persia. From many parts of western Europe these splendid animals have, however, been exterminated; and in the British Isles they survive in a wild state only in Devon and Somerset, the highlands and isles of Scotland, and parts of Ireland. The red deer of the Caspian district and the neighbouring countries, commonly known (from its Persian name) as the maral, is a much larger and also a greyer animal, with heavier antlers, than its west European representative. The latter have been split up into several local races, which need not, however, be particularised in this place.

The food of the red deer varies considerably according to the time of year, and comprises grass and other herbage, corn, leaves and boughs, bark, acorns, chestnuts, funguses, lichens, and moss. In autumn, when living near cultivated ground, deer will dig up with their hoofs, potatoes, artichokes, and other edible roots.

The pairing-time commences early in September, and continues till the middle of October; and at this season, when they utter the well-known bellowing or roaring, the stags not only fight fiercely among themselves for the mastery of the herd, but are highly dangerous to human beings. At no time very amiable, the stags at this season are little better than incarnate fiends. Soon after the breeding-season the antlers are shed, to be replaced by new growths, covered at first with soft velvety skin, the following spring.

At the end of May or early in June the hind seeks a sequestered situation amid covert in which to give birth to her fawn. The fawns, of which there may occasionally be twins, are extremely helpless at birth, but in a short time gain sufficient strength to run by the side of their mothers.

THE BEAVER

(Castor fiber)

THE beaver enjoys the distinction of being the only warm-blooded quadruped that is in the habit of making really noticeable modifications in the appearance of the earth's surface. Many quadrupeds, such as foxes, ant-bears, rabbits, and rats and mice burrow holes in the ground, while the mole marks the course of its subterranean tunnels by throwing up heaps of earth at intervals. But although such excavations and hillocks, when sufficiently numerous, may to a slight degree affect the appearance of a meadow, they are nothing in comparison to the changes brought in a valley by a colony of beavers. By throwing a dam across its course, these industrious rodents will convert a narrow stream into a wide sheet of stagnant water, which in the course of time may become silted up so as to form a broad and level "beaver-meadow," where there was originally a rocky valley. In or near their dams beavers likewise construct dwellings of mud and clay, known in America as "lodges," for their own accommodation.

But this is by no means all beavers accomplish in the way of "public works," for, by means of the single pair of powerful chisel-like teeth in the fore part of each jaw and the powerful muscles by which the jaws themselves are worked, these animals, which are about the size of an ordinary spaniel, are enabled to fell trees of considerable size, which are used in the construction of their dams.

Beavers are the sole living representatives of a family of rodents allied on the one hand to squirrels and dormice, and on the other to rats and mice. Two structural peculiarities are very characteristic of these rodents. In the first place, one of the toes of the fore-foot is provided with a double claw, which may be used in dressing the beautiful, long brown fur; a similar structure occurring in the smaller rodent known as the Arctic lemming (*Dicrostonyx torquatus*). Secondly, there is the remarkable flattened, scaly tail, which almost looks as though it did not belong to the animal, although in reality, except for its superior size, it is not much more abnormal than the scaly cylindrical tail of the rat. Several myths attach to the beaver's tail; it was said, for instance, to be employed as a trowel for plastering down the mud used in building the dams and lodges, although its real use is to act as a rudder in swimming, more especially when its owner is transporting the trunk of a felled tree. When entering the water, or when engaged in playing therein, the beaver frequently makes a resounding "smack" by striking the surface with its tail.

In monkish times beaver-tail was considered to partake more of the nature of fish than flesh, and was consequently allowed to be eaten on fast days. This was, however, in the days when beavers were still abundant in all the great rivers of Europe, from most of which they have now been all but exterminated for the sake of their valuable fur, and likewise for the odoriferous secretion known as castoreum, which was formerly much used both in medicine and in perfumery. When the last beaver was killed in the British Isles is unknown, but the species still survived in Wales when the old chronicles were written; and we have testimony as to its former existence in England not only in the shape of skulls, teeth, and bones dug up from time to time in the peat of the fens and other superficial deposits, but also in place-names such as Beverley, in Yorkshire.

Considerable colonies of beavers still exist, by the aid of special protection, in certain parts of Scandinavia, while a few are taken from time to time in the Rhone, but from the Rhine, and even the Vistula, they seem to have completely disappeared. In eastern Russia they probably still survive locally, as they doubtless do over a large part of Siberia, although our information on this point is very defective. Indeed the southern range of the beaver in central Asia seems to be still unknown, although it is certain that the species never existed in Kashmir or the Himalaya.

Of late years it has been suggested that each of the great European river-systems possessed a special race of beavers of its own; but the evidence adduced in favour of this opinion is at present insufficient. Speaking broadly, the beaver may be regarded as a circumpolar animal; although its American representative has been separated, on account of a comparatively small difference in the shape of the bones covering the cavity of the nose, as a distinct species, under the name of *Castor canadensis*. Unfortunately, the Canadian beaver has been almost as much persecuted as its European relative, and has been exterminated from many districts.

Beavers, it need scarcely be mentioned, are thoroughly aquatic rodents, which feed on vegetable substances, and have their entrances to their habitations under water. They remain active all the winter, when they swim beneath the ice. In Europe beavers have given up constructing lodges, and live in burrows.

THE MARMOT

(Arctomys marmotta)

THE marmot of the Alps is the typical representative of a large assemblage of burrowing rodents near akin to squirrels; the head-quarters of the group being in central Asia. By rights, of course, the name belongs exclusively to the typical species, but as it has been extended to include all the members of the group, the former is now distinguished as the Alpine marmot. Marmotte, it appears, is the Savoyard name of this rodent, which in the Engadine is designated marmotella, while its German designation is murmeltier.

The typical marmot is confined to Europe, and mainly to the high ranges of the Alps, Pyrenees, and Carpathians. In eastern Europe and western Asia it is replaced by the bobac (*Arctomys bobac*); and in central Asia there are a number of species, of which several are considerably larger and more brightly coloured than the Alpine animal; none is found to the south of the main range of the Himalaya, and the group is represented elsewhere only by the so-called woodchuck (*A. monax*) of North America, which attains a length of about 24 inches.

The general colour and appearance of the Alpine marmot are well shown in the Plate; and it will suffice to direct attention to the shortness of the ears as a feature connected with burrowing habits. In one Himalayan species the tail is considerably longer. In the Alps marmots dwell high up among bare rocks, above the zone of vegetation, where not even goats venture, and where it frequently snows for six weeks together in winter. In such desolate situations these hardy rodents make their home in the little islands of rocks between the glaciers. Himalayan marmots, on the other hand, live at an elevation where a considerable amount of vegetable growth flourishes in summer; their burrows being frequently excavated beneath clumps of wild rhubarb.

As summer resorts, Alpine marmots select situations with a southerly, easterly, or westerly aspect on the mountain slopes, and here they construct their summer dwellings, which are designed to afford them shelter in bad weather and to serve as a refuge from danger. In autumn they dig deeper into the sides of the mountains to construct their winter burrows and chambers, which have to afford accommodation for the entire family, whose number generally ranges between five and fifteen individuals. The burrow terminates in a large chamber, which is filled with soft, short hay. As early as August these rodents begin to collect grass, which is spread out on the hillsides to dry, and then carried into the burrows.

The entrance to the burrow is only just wide enough to admit the owners, and at the commencement of the winter-sleep is blocked with earth, stones, clay, grass, and hay. When this work is completed, the whole family falls into a death-like sleep lasting from six to eight months; in this torpid condition, well protected from the intense cold, they remain till awakened to new life by the warmth of returning spring.

Marmots feed upon a number of different kinds of Alpine plants, as well as on the fresh or dried grass found in the neighbourhood of their burrows. When eating, they sit up on their hind-legs and hold the food in their mouths after the manner of squirrels; and the herbage they consume is so full of sap, that they seldom require to drink. The same upright position is assumed when they first come out of their burrows, in order that they may see whether all is safe; if danger threatens, they utter a shrill whistle and disappear with marvellous rapidity into the depths of the burrow. Frequently they take up their position on some large rock in the neighbourhood of the burrow, on which they can either sit up and survey the prospect, or bask in the warmth of the summer sun.

Their extraordinary wariness and the rapidity with which they disappear from view are due to their numerous enemies, among which man, foxes, and eagles and other birds-of-prey are the chief. It requires only the shadow of an eagle's wings to make them utter their piercing, whistling scream and vanish into the holes, from which they do not reappear for some time. As a rule, they only remain above ground while the sun is shining, and they keep entirely below during bad weather, so that they are regarded by the peasants as weather-prophets.

To the Savoyards and other Alpine peasantry the marmot is a valuable animal; its flesh being much esteemed as food, when it has been freed from a certain disagreeable odour by smoking. The fat is regarded as a remedy for many diseases; and a freshly removed marmot skin is considered wonderfully efficacious in cases of rheumatism. Marmots are caught either by trapping or by digging them out of their burrows. Years ago Savoyard organmen used frequently to be accompanied by a marmot or two.

THE HAMSTER

(Cricetus frumentarius)

THE hamster, which, although abundant in many parts of the Continent is unknown in the British Isles, is the typical representative of a large section of the mouse tribe characterised by the cusps on the upper cheek-teeth forming two longitudinal rows, instead of the three found in those of ordinary rats and mice. In size it may be compared to a rat, but its tail is reduced to a mere stump, not more than a couple of inches in length; while it is further characterised by the brilliant and variegated colouring of its fur. Short-legged and stoutly built, it has relatively small, membranous ears, large brilliant eyes, a rather sharp muzzle, small toes, and short claws. The glossy, hairy fur is underlain by a thick woolly under-fur. Very characteristic of this animal is a narrow line of fur darker than the rest on the middle line of the back which marks the position of a gland.

In general colour the fur of the upper-parts is light brownish yellow; but the sides of the face are variegated with chestnut and white, and there is a white area on the shoulder, while the under-parts and the greater portion of the limbs are black, the black extending upwards to some extent behind the fore-legs. Hamsters are, however, subject to considerable individual variation in colour, and black, pied, and even white examples are by no means uncommon.

Soft dry soils other than loose sand, which is unsuited for its burrows, form the favourite resorts of the hamster, which in localities of this nature ranges from the valley of the Rhine to that of the Obi in Siberia. Its distribution is, however, very local, and the species is unknown in the southern and south-western districts of Germany, as it is also in eastern and western Prussia: in Thuringia and Saxony, on the other hand, it is abundant.

The chief interest of the hamster is concentrated in its remarkable habits. These rodents associate in large societies; and, like marmots, construct both summer and winter burrows, in the latter of which they become torpid during the cold months of the year. The winter-burrow includes a large sleeping-chamber, situated at a depth of from one to two yards below the surface; and also a storehouse or granary, in which quantities of corn of various kinds are collected by these industrious rodents for use during such portions of their retirement as they are active. The burrow leading to the dwelling-chamber descends almost perpendicularly, but takes a turn before opening into the chamber itself, which is likewise provided with an oblique emergency exit. Although grain forms their chief nutriment during the period of retirement, hamsters in summer consume large quantities of peas, beans, roots, fruits, and grass and other green food.

As a rule, hamsters retire from the world to their subterranean dwelling-places some time during October; when they block up the entrances and exits of the winter-burrows with earth. Apparently they almost immediately enter on their winter-sleep, from which they do not awake till the following February or March, according to the temperature. The weather at this early period of the year is, however, by no means suited for an out-door existence, and these rodents accordingly subsist for a season on their hoarded grain. The old males generally make their appearance above ground about the middle of March, but the females defer their emergence till a fortnight or so later. They are then ravenously hungry, and will devour almost anything that comes in their way, including beetles or other insects and an occasional bird or mouse.

In summer the nest-chamber of the females, which is distinct from the summer-burrow of the males and is furnished with one exit and several entrances, is carefully lined with hay. Towards the end of April the males visit the females in their own apartments; and four or five weeks later the first litter of blind and naked young—varying in number from half a dozen to eighteen—is produced. These rapidly develop their fur, and open their eyes about the eighth or ninth day; and within a fortnight are driven away from the parental burrow to construct a new one of their own. Freed from one family, the old hamsters set about producing a second one, which usually comes into the world in July. The annual increase is, however, by no means limited to the older individuals, for the members of the early spring litter are able to produce young ones in the autumn.

Hamsters frequently make their appearance in enormous swarms, when they do vast damage to crops. In return, the winter granaries of these rodents are frequently raided by the peasants of countries where they are common; the flesh of the hamster is also eaten, and its fur employed for lining cloaks and coats.

THE DUCKBILL or PLATYPUS

(Ornithorhynchus anatinus)

FOR many years it was reported by the natives of Australia that the extraordinary warm-blooded quadruped known to naturalists as the duckbill, or platypus, produced its young from eggs laid in a burrow by the female. That a mammal—and a mammal, although of an altogether peculiar and out of the way type, the creature undoubtedly is—should lay eggs was, however, too much for the minds of stay-at-home naturalists, and the fiat accordingly went forth that the native story was to be discredited. And discredited it therefore was. In nature, as in other things, truth will, however, ultimately prevail; and we now know for certain that the female lays in a burrow in the bank of some river or pool a couple of hard-shelled, oval eggs, which in due course hatch out into naked, helpless young, furnished with soft sucking lips. Not that they suck in the ordinary mammalian fashion, for the female platypus has no nipples, but her milk oozes out in the breast from a number of sieve-like pores, from the surface of which it is sucked up by her offspring.

Such a difference from the ordinary mammalian way of doing business proclaims the wide distinction between the platypus, and, it may be added, its relatives the spiny ant-eaters or echidnas (one of which forms the subject of another illustration), and all other warm-blooded quadrupeds. Nor is this all, for in the structure of their skeleton and soft internal parts the platypus and the echidna display many marks of affinity with reptiles and birds, which are totally wanting in other mammals. These two creatures represent indeed a group by themselves, so that mammals may be divided into two great primary sections, the one embracing only the two egg-laying types, and the other all the rest.

And it is not a little significant that the egg-layers are confined to Australia and New Guinea, the home of many other primitive and ancient types which have disappeared from the rest of the world. In one sense indeed the platypus and the echidna are not exactly primitive creatures, as they have several specialised characters which were evidently wanting in their ancestors. They may rather be described as specialised branches of an ancient and primitive stock.

The duckbill is a heavily made aquatic mammal of about the size of a very short-legged rabbit, with blackish, mole-like fur above, passing gradually into whitish beneath, and a short, thick, tapering tail. The very short limbs terminate in thick toes, connected together by a web and armed with strong, pointed claws. In the fore-feet the margin of the web projects considerably beyond the claws, but on the rare occasions that the animal leaves the water the margin is folded downward beneath the claws, so as to leave the claws exposed. In museum-specimens, however, the web is almost invariably shown fully expanded, as in the accompanying illustration; a condition in which it would obviously be impossible for the animal to walk on land.

The most remarkable external feature of the duckbill is undoubtedly the duck-like, naked beak, pierced with two holes representing the nostrils. In stuffed specimens, at any rate, this beak is dark-coloured and horny in consistence, and looks as though it did not belong to the animal, but in life it is soft and tender. Medium-sized, dark eyes complete the physiognomy of this strange creature, in which external ears are wanting.

The internal arrangements of the mouth of the duckbill are scarcely less curious than the exterior. In early life the mouth is furnished, both above and below, with three pairs of somewhat quadrangular cheek-teeth, with raised and cusped margins. Beneath these grow up certain large horny plates, and about the time that full maturity is attained the teeth become worn out, and are finally shed, thus leaving the horny plates as the sole masticating organs.

This replacement of the teeth by horny plates appears to be connected with the nature of the food, for while in early life the duckbill appears to subsist mainly on water insects and other comparatively soft aquatic creatures, later on it takes to feeding almost entirely on bivalve shells of one particular species; and for crushing the stout shells of these molluscs it has been suggested that the tough horny plates are better suited than brittle teeth.

Duckbills, except when in their burrows, pass the greater portion of their time in the water, selecting quiet pools for their favourite haunts. In such situations they may be seen on a still evening floating and diving, and looking more like bottles in the water than anything else. They obtain their food chiefly by probing in the mud with their duck-like beaks. The dwelling-chamber of the burrow is situated in the bank above the water level, but its entrance is below the surface, although there is also an exit on the land. In the pairing-season the males, which are armed with a poison-bearing spine on the inside of each hind-leg, fight fiercely among themselves.

The duckbill, of which there is but a single species, is absolutely confined to southern and eastern Australia and Tasmania; and its nearest living relative is the echidna, of which a picture and notice follow.

THE SPINY ANT-EATER OR ECHIDNA

(Tachyglossus aculeatus)

NO one who looked at the portrait of the spiny ant-eater for the first time, and had no knowledge of its anatomy or history, would be likely to guess that it was a near relative of the duckbill. But in natural history, when we have to deal with members of different groups, externals count for very little, and all depends upon internal organisation. In the latter respect the echidnas, for there is more than one species, resemble in all essential features the duckbill, as they do in laying hard-shelled eggs, from which the young are eventually hatched. The single egg of the echidna, in place of being laid in a burrow, is, however, carried about by the female in a pouch developed for the purpose on the under side of her body shortly before the egg is laid; and in this same temporary pouch the young is likewise nurtured during the earlier stages of its existence.

The duckbill and the echidna afford an excellent example of the diversity of appearance produced in animals more or less nearly related to each other by specialisation and adaptation to totally distinct modes of life. In the duckbill the specialisation and adaptation are for an aquatic existence; in the echidna they are for a burrowing, terrestrial life and a diet of ants.

To an ant-eater teeth of any kind would be not only useless, but an actual hindrance, and they have accordingly been discarded, while the muzzle has been prolonged into a decidedly bird-like beak. In this respect the echidna much resembles the great South American ant-eater, which belongs to a totally different group of mammals.

To enable it to dig out the nests of the ants which form its chief food, and likewise to excavate the burrows in which it passes the day, the echidna is armed with powerful claws, those on the hind-feet being, however, much larger and more curved than those in front. It is with these strong hind-claws that the earth loosened by the fore-feet is thrown out from ants' nests and the burrow. Like the porcupine, which is also a nocturnal and a burrowing creature, the echidna has its back protected with an array of parti-coloured horny spines mingled with hairs. The degree of development of the spines is, however, subject to great variation; and there is one race in which the hair predominates, and the spines appear only in the midst of the dense brown fur. Like the platypus, the echidna has no external ears.

The ordinary, or five-toed, echidna has a much more extensive range than the platypus, occurring, in suitable localities, not only all over Australia and Tasmania, but likewise in New Guinea. The last-named island is likewise the home of the much larger three-toed echidna (*Proechidna bruijnii*), in which the beak is longer, more slender, and distinctly curved, while the number of the toes on each foot is reduced to three.

With the commencement of evening the echidna issues forth from the lair or burrow, in which it has passed the day, in search of food—this comprising not only ants, both ordinary and white, but likewise such other insects and their grubs as may be encountered or dug up during the nocturnal wanderings. Soon after daybreak the creature returns to its burrow. During the hottest and driest part of the Australian summer spiny ant-eaters fall into a torpid condition, when they exist for weeks at a time on no other nourishment but their own fat. In cases of extreme hunger they are stated to fill their stomachs with sand. At the end of the dry season, when rain falls and the country resumes its verdure, the echidnas wake up, and the males relinquish their normally solitary life and take to themselves partners.

On occasions of danger the echidna has two means of defence—it can either roll itself up into a ball and present a sphere of spikes to its enemy, or it can burrow with such rapidity that it actually seems to sink into the earth as if swimming.

The single egg appears to be conveyed to the pouch by the female in her mouth, and the parent assists the young echidna, whose muzzle is armed with a special knob for that purpose, in breaking the shell. Naked and blind when first hatched, the young one remains in the pouch till its spines make their appearance, drawing its nutriment from two pores through which the milk flows. When the young echidna has been turned adrift in the world to shift for itself, the maternal breeding-pouch shrivels up, to be re-developed the following year.

As in the case of the platypus, no remains of extinct echidnas are found anywhere except in the superficial deposits of Australia itself. Certain teeth from rocks older than the Chalk, both in Europe and America, may, however, indicate the ancestral stock of the egg-laying group.

THE BLACK SWAN

(Cygnus atratus)

WHEN the old Roman poet penned the well-known line, "Rara avis in terris, nigroque simillima cygno,"[A] he little imagined that a black swan was actually living in the then unknown and undreamt-of Antipodes, to be discovered in the dim and distant future by the sailors of a little island in the far north inhabited by fair-haired barbarians of whom he may or may not have heard.

[A] "A rare bird in the world, and most like a black swan."

The black swan of Australia, from which the Swan River takes its name, is not perhaps the ideal black swan of the poet, for, as a matter of fact, it has a good deal of white in its wings, although but little of this is visible when the wings are closed, as in the illustration. Moreover, the general colour of the plumage is brownish black, rather than pure black, while the beak is crimson with a white tip, and the eyes are scarlet. From all other swans this species differs by the curling feathers in the region of the shoulder, and the extreme shortness of the tail. Other distinctive features are to be found in the compressed body, the long neck, the small, gracefully carried head, and the absence of a ridge to the beak.

The range of the black swan includes both Australia and Tasmania, where it is found not only on the coast, but likewise on the rivers and lakes of the interior. In the earlier days of Australian colonisation it abounded in many parts of the country; but, unfortunately, it is a bird by no means shy, and therefore comparatively easy to shoot, and in consequence of this it has been incessantly and relentlessly persecuted, even the cygnets, while still unable to fly, being hunted among the reeds in the swamps and killed out of pure maliciousness. As a result of this persecution the species has everywhere become scarce, and from some parts of the country has been exterminated.

During the winter months these swans collect in small parties and families in South Australia, and return to their breeding-places in spring. The nest is a huge, ill-built structure, made out of coarse materials and lined with sedge. It is generally situated in the neighbourhood of small islands, where the parent birds can readily collect the large amount of material required for its construction. The clutch consists of from five to seven dirty white or pale green spotted eggs, which are brooded with great assiduity by the female, while the male keeps guard in the neighbourhood. The cygnets are clothed in a grey or rufous down, and can swim and dive as soon as they are hatched. The food of both young and adults comprises water-plants of all kinds, as well as worms, molluscs, and small frogs and fishes.

The black swan is a noisy species, more especially during the pairing-season, when it gives vent to loud trumpetings. When doing so, the neck is stretched out straight along the water, so that the beak lies close to the surface. At this season the males become very pugnacious, and fight much among themselves.

It is a beautiful sight to watch a party of these swans swimming on a still lake or river by moonlight in large circles; but this is exceeded when the birds mount into the air and reveal the striking contrast formed by the white of the pinions against the sable of the rest of the plumage. When flying, these swans stretch out their necks to the fullest extent, and accompany the loud and regular beat of their wings with continuous and resounding trumpetings.

The black swan bears captivity well, alike in its own country and in Europe, being content with small rations and breeding regularly every season.

The Australian black swan is by no means the only abnormally coloured member of its tribe inhabiting the Southern Hemisphere, its place in South America being taken by the black-necked swan (*C. nigricollis*), which is a native of Chile, Argentina, and some of the other southern countries. In this handsome species the head and neck are black, and the remainder of the plumage white; the lores, or bare patches in front of the eyes, and the basal portion of the beak being red. In its long and wedge-shaped tail the black-necked swan differs from the black species and resembles the ordinary

European mute swan (*C. olor*), from which it is structurally distinguished by the scalloped margins of the webs of the toes. The adults of all swans, except the black and the black-necked species, are wholly white; the smallest kind is the coscoroba swan of South America, which is no larger than a goose, and has been referred to a genus by itself.

THE BUSTARD

(Otis tarda)

BY the extermination of the bustard, or great bustard, as it is sometimes called in order to distinguish it from its smaller relatives, the British Islands have lost one of the finest members of their bird fauna; and, unfortunately, all attempts to rehabilitate this magnificent species have resulted in failure. It is to be feared, moreover, that any such attempts have but little chance of success in the future, for the bustard is a native of open downs and fallows, where it must live an undisturbed and untrammelled existence, and in England at the present day this is almost an impossibility in country of that description. In this respect the bustard stands at a great disadvantage in comparison with the capercaillie, whose reintroduction into the forests of Scotland was a relatively easy matter.

Like those of most polygamous birds, the cocks of the bustard are much larger than the hens, rivalling full-grown turkeys in the matter of size; they are also much more attractively coloured, and are furnished with a quantity of white fluffy plumes, which are only fully displayed when the birds perform their curious nuptial parade to attract the hens, although some of them may be seen when two cocks are fighting, as depicted in the plate. Another peculiarity of the cock is the possession of a great pouch, communicating with the windpipe, on the throat, which can be inflated under the influence of excitement.

When the wings are closed and the bird is engaged in feeding or other normal occupation, the colouring of the body-feathers of the cock bustard is admirably adapted to harmonise with the generally sandy or earthy hue of the surroundings. In the case of the hen the whole plumage is protectively covered. The one feature of the body-plumage in both sexes which produces such a wonderful harmony between the colouring and that of the surroundings is the presence of a vast number of narrow black bars on a rufous buff ground, this type of colouring not only matching sandy or loamy soil, but likewise aiding to break up the outline of the bird.

When bustards lived in England their favourite resorts were the dry, heathy uplands of Norfolk and Suffolk, and the downland of Cambridgeshire, and the neighbourhood of Royston. At the present day these splendid birds are still common in many parts of the Continent, their range including the greater portion of central and southern Europe and a large extent of central Asia, while in winter it likewise embraces northern Africa. The steppes of Russia, the plains of the Danube in its course through Hungary, and the open tracts of central Asia are the regions where bustards are now to be met with in the greatest numbers, although large flocks may be seen in parts of Spain.

Bustards, which associate in large flocks, are essentially birds of the open country, never entering woods, and preferring elevated ground, whence they can command a wide view of the plains below, to which they descend during the daytime in search of food. They are exceedingly shy and mistrustful, giving every bush in their path a wide berth, lest it should conceal a lurking foe, and taking to flight on the least alarm. For warning of the presence of danger they seem to depend mainly upon sight, although their hearing is also good; on the other hand, their sense of smell, like that of most birds is but poorly developed. This excessive shyness renders the bustard a difficult bird to bag, even by the experienced gunner.

At the commencement of the pairing-season the cocks endeavour to attract the attention of the hens by their nuptial display. Advancing towards the latter they ruffle out their feathers like an excited turkey-cock, at the same time lowering their wings and spreading out the tail like a fan. The gular pouch is also inflated, so as to make the neck appear as thick as possible, and the beautiful under feathers are displayed to the utmost extent, the body and neck being half buried in a billowy mass of snow-white plumes. It is at this time also that the cocks rush at one another with trailing wings to contend with beak and talons for the mastery. On such occasions it is possible to capture these generally shy birds with the hand.

The nest is generally constructed amid growing corn, the hen merely scratching a slight hollow in the ground, which she lines with straw, bents, and grass. The three or four eggs are incubated by the hen alone, and when the young are hatched, both these and their parents leave the standing corn only for short periods. At first the young bustards feed chiefly upon insects and worms, to which they are directed by the old birds, but later on vegetable substances constitute almost their sole food. Young peas and cabbage form very favourite food, but in default of better nutriment grass is largely eaten.

THE SOMALI OSTRICH

(Struthio molybdophanes)

OSTRICHES, which are natives of the deserts and semi-desert plains of Africa and south-western Asia, cannot possibly be mistaken for any other birds, and therefore stand in no need of description, although it may be as well to mention that they are the largest of living birds, and are absolutely peculiar in having only two toes to each foot, as well as in the absence of feathers on the thigh. It was at one time considered that the lack of the power of flight, which characterises these birds in common with their relatives the rheas of South America, the emu of Australia, the cassowaries of Austro-Malaya and Australasia, and the tiny kiwis and rheas of New Zealand, was a primitive feature. But this is manifestly an erroneous idea, and all these groups are evidently descended from birds endowed with the power of flight, their nearest relatives being the tinamus of South America and the game-birds, with which they agree in the presence of longitudinal light stripes on the downy dress of the chicks.

Naturalists now recognise four kinds of African ostriches, which, although generally classed as distinct species, might perhaps be better regarded as local races of a single specific type. These, regarded as species, comprise the typical red-legged and red-thighed North African ostrich (*Struthio camelus*), ranging into Palestine and Arabia, and laying thickly pitted eggs; the Somali ostrich (*S. molybdophanes*), characterised by its bluish grey neck and thighs; the Masai ostrich (*S. masaicus*) of East Africa, with the bare parts red, the body-plumage of the cocks brownish black in place of black, and the eggs pitted; and, lastly, the South African *S. australis*, which lays smooth-shelled eggs and has the neck and thighs light bluish grey. In the Somali ostrich, which inhabits Somaliland and western Gallaland as far as the Juba River, the colour of the bare parts of the body may best be described as grey or slaty blue, while the margins of the beak and the front surfaces of the lower part of the legs are dull vermilion.

Ostriches are essentially gregarious birds, associating either in small family parties comprising five or six birds, or in large flocks, which in East Africa mingle freely with herds of hartebeests, gnus, and bontequaggas. They require a wide extent of open country, with grass and, above all, water, of which they drink frequently and copiously. Although ostriches feed chiefly upon vegetable substances, they likewise consume worms, insects, molluscs, reptiles, and probably also small birds and mammals, while in captivity they will swallow almost anything that is offered them, including nails, keys, and copper coins. In a state of nature they swallow sand, earth, and small stones, in order to assist the action of their gizzards.

The cocks appear to be polygamous, although some doubt has been expressed as to whether this is really the case; and it seems certain that several hens lay in the same nest, where as many as thirty eggs may be laid in a slight hollow in the sand. Incubation is undertaken by the cock alone, and it is he who looks after the chicks, which he tends with remarkable care. The chicks when they escape from the eggs (each of which weighs as much as twenty-four hens' eggs) are as large as hens; and, with their bristly feather-quills, are rather suggestive of two-legged hedgehogs.

Ostriches are extremely shy and remarkably swift birds, so that they are difficult either to stalk or to ride down. When brought to bay, or when running loose in captivity, they should be approached with great caution, as a kick from their strongly clawed feet will prove fatal to an animal of the size of a jackal, if not also to a human being. The vision of these giant birds is very strongly developed and capable of sweeping the country to the distance of at least a mile; and their senses of hearing and smell are also good, although taste appears to be practically wanting. Occasionally ostriches indulge in a kind of dance; and they are also subject, when startled, to sudden collapse, which may have given rise to the old fable of their burying their heads in the sand and leaving their bodies exposed.

If allowed sufficient space, ostriches thrive remarkably well in confinement; and ostrich-farming, for the sake of the lovely white plumes of the cocks, has become an important industry, not only in Africa, but in many other parts of the world. Ostrich-farming was commenced in Algeria, and by 1865 had been introduced in Cape Colony, whence it has spread northwards, while it has also been established in Egypt, southern Russia, California, Patagonia, and Argentina. The cocks are clipped regularly every eight months.

THE PIED HORNBILL

(Buceros bicornis)

THE hornbills, which are as large as big eagles, are some of the most extraordinary and bizarre of all birds, not only on account of the huge beaks from which they take their name, but from their strange and unique nesting-habits. In the enormous development of the beak these Old World birds are paralleled by the toucans of tropical America, although the two groups are in nowise nearly related. Such a huge and apparently unnecessary structure might at first sight seem too heavy and bulky for the bird by which it is carried; but internally the beak is generally a mere delicate tissue of bone, so that its weight is really very slight. Indeed, these enormous beaks are in most cases probably much less of an incumbrance to their owners than are the matinee hats favoured by ladies of the present day.

The pied, or double-horned species, the hornray of the natives, inhabits the dense forests of the Indo-Malay countries from the Himalaya and the Indian peninsula to the south of China and Sumatra. The glossy plumage is mostly black above, with the under-parts, the neck, the tips of the flight feathers, and the tail feathers more or less white or whitish, the upper half of the huge beak, with its horn-like appendage, being reddish yellow, and the lower half yellow. The total length of the bird, which is the largest member of the whole group, exceeds a yard. The so-called "horn," or casque, forms a more or less flat and oblong platform on the upper part of the beak and the fore part of the head, terminating in front in two obtuse corners. On the other hand, the same appendage in the nearly allied rhinoceros-hornbill, or engan (*Buceros rhinoceros*), is much more horn-like, as it turns up into a recurved point in front.

The range of the hornbill group includes Africa, India, Malaya, Celebes, and New Guinea and the neighbouring islands as far as the Solomon group. The ground-hornbills are solely African, while the hollow-casqued section has a distribution equivalent to that of the group. One very remarkable Malay species, the solid-billed hornbill, the teban-mertua of the Malays (*Rhinoplax vigil*), differs from all the rest in having the beak and casque solid. This species is much hunted by the Chinese for the sake of its beak and casque, which have a structure resembling ivory, and are used as a substitute for that material in carving; while it is further remarkable for its bare neck and the elongation of the middle pair of tail feathers.

The flight of hornbills is heavy and noisy, the swishing sound produced by the wings being audible at a great distance. The whole body of these birds is permeated by air-tubes connected with the wind-pipe, and it is said that the movement of the air in these increases the noisiness of the flight.

Mountain-forests bordering valleys, and big forests near rivers, are the favourite haunts of the typical hornbills. Here the great birds perch in flocks which may number a score in individuals; for hours together they sit motionless, with the neck drawn back on to the body, and the body itself pressed down on to the feet.

Their food consists chiefly of the fruits of the trees in which these birds take up their quarters, the favourite fruit being figs; but in captivity hornbills will eat fruits of all kinds, as well as animal food. These birds kill every small creature that comes in the way, and throw them up in the air before swallowing the body. When perching, hornbills utter from time to time a loud cry, recalling the bark of a dog, this changing into a loud scream when they are wounded.

The most remarkable feature connected with the habits of these birds is, however, connected with their nesting arrangements. As was first observed in India in 1855 by a military officer devoted to outdoor natural history, when the female has laid her clutch of eggs, varying in number from two to half a dozen, in some hole in a tree and commenced to incubate, the male walls up the entrance to the apartment with clay, leaving an aperture only just large enough for her to protrude her beak and receive the food brought regularly by her hard-working mate. The object of this immuring process is doubtless to protect the female and her eggs or young from the attacks of birds-of-prey, or predatory arboreal mammals. Under these circumstances sanitary matters cannot, of course, be attended to, and the nest-hole, consequently, soon becomes an evil-smelling mass of abominable filth. In order that no time may be wasted, the female moults during the period of her confinement, a process which adds still more to the mess in the hole. The young, which come forth from the eggs almost naked, remain in the breeding-hole till they are fully fledged. To reach the food brought by her partner, the female, at any rate during the early stages of incubation, has to climb up to the aperture in the wall of clay.

THE BLUE MACAW

(Ara cærulea or *Ara ararauna)*

THE blue, or blue and yellow, macaw, the ararauna of the natives, is one of the most familiar representatives of a group of tropical American parrots characterised by their large size, long tails, and brilliantly coloured plumage, as well as, it may be added, by their atrociously loud, screaming cries, which render them among the most objectionable of all birds as pets. In the species forming the subject of the plate blue and yellow are the predominant hues; yellow, it is said, being the equivalent among true parrots, in which white is unknown, to no colour at all. The whole of the under surface and the sides of the neck are deep orange-yellow, while the upper surface, inclusive of the tail-coverts, is dark sky-blue. The range of this species, like that of its relative the green and blue macaw (*Ara severa*), extends only from Panama southwards through the tropical forests of the New World. Other kinds are, however, found farther north, the handsome red and blue species (*A. chloroptera*) and the red and green macaw (*A. militaris*) extending, for instance, from Mexico and central America to Bolivia. Four other species, known on account of their deep blue colouring, as hyacinthine macaws, constituting the genera *Anodorhynchus* and *Cyanopsittacus*, are, on the other hand, solely Brazilian.

Macaws are not only strong fliers, but likewise excellent climbers, and in the latter mode of progression make use of their powerful beaks as well as their feet, in captivity, at any rate, frequently hanging from a branch or perch by the beak alone, which is specially adapted for obtaining a hold on smooth branches by the sharp downward curvature of the extremity of its upper half. The food of these birds includes fruits and seeds, especially hard nuts of various kinds, which are cracked in the strong beak as if in a vice. When the fruits or nuts are large, they are held to the beak in one foot. The thick, fleshy tongue aids in extracting the inside of fruits from the rind, or the kernels of nuts from their shells.

All macaws are essentially birds of the great primeval forests of tropical America, the blue and yellow species being especially abundant in those of northern Brazil. During the hottest hours of the day these birds sit quietly perched on the lower branches of thick-foliaged trees, the long tail hanging straight down, and the neck being drawn in. After resting for several hours, they issue forth in search of food; and a flock of these magnificent birds on the wing in the clear air of the tropics is a truly splendid sight, as they fly slowly to and from their feeding-places. Having selected a likely looking tree, the whole flock descend with a rush, and proceed to strip it of its fruit. Quickly each bird climbs out to the end of a branch and sets to work at cracking nuts, or devouring softer fruit. When such a flock of araraunas is feasting, little is seen of the birds themselves, but their presence is amply manifested by their incessant and discordant cries and screams. If a shot be fired in the neighbourhood, the whole flock rises immediately in a gold and azure cloud. Macaws of all kinds constantly make raids on cultivated crops, where, unless promptly driven off, they do incalculable damage in a surprisingly short space of time, cocoa being one of the crops to which they are specially partial.

As is the case with all members of the parrot tribe, the union between the sexes of the araraúna is very close: the two members of a pair are thorough comrades, and live only for themselves and their offspring. At the pairing-season each couple of the older birds resorts to the same spot, and often to the same tree, as has served them previously as a breeding-place. For the nest a tall tree is always selected, and a slit in the trunk or a hollow in one of the branches is enlarged by means of the powerful beak until a cavity large enough to contain the nest is excavated. In this chamber the female ararauna deposits a couple of white eggs of nearly the same size as those of a domesticated hen, afterwards incubating them with great assiduity, her long tail projecting all the time out of the hole, and thus revealing the situation of the nest.

Tame araraunas and other macaws are much esteemed by the Indians of Brazil and other South American countries. These dwell on the properties of their owners, and fly off to feed in the fields, just in the same manner as domesticated pigeons in Europe. They form the ornaments of the Indian poultry-yard.

In captivity in Europe macaws thrive well, and can be taught to speak, although they never turn out such good linguists as many other members of the parrot tribe.

THE MANDARIN DUCK

(Æx galerita)

THE male of the mandarin duck—the mandarin drake, as it should properly be called—looks so essentially Chinese that one would almost intuitively guess that it came from China. Nevertheless, it has a near relative, and the only other member of the same genus, in North America, where it is represented by the summer duck (*Æx sponsa*). In this distribution these painted ducks, as they have not inaptly been called, resemble several other groups of animals, such as the true alligators (*Alligator*) and the spoonbeaked sturgeons (*Polyodon* and *Psephurus*), each of which has one North American and one Chinese species, although the American and Asiatic spoonbeaked sturgeons are referred to separate genera.

Ducks, it may be mentioned, are divided into two distinct sections—namely, the swimming ducks and the diving ducks; the mandarin duck and its American relative belonging to the former section, of which they form a special group, characterised by the ornate plumage of the drakes, and particularly by the broad ornamental feathers on the shoulders and the elongated plumes on the head, which form a kind of helmet, with a large flange extending backwards over the neck.

The mandarin drake is specially distinguished from the male of the summer duck by possessing a ruff on the neck formed of elongated, narrow, chestnut feathers streaked with whitish, and a chestnut and black "fan" rising up in a kind of tiger's-claw shape on each side of the hind end of the body, and mainly formed by the broad innermost secondary feather of the wing; the helmet being copper-coloured, purple, and green, and separated by a broad white band enclosing the eye from the chestnut throat-ruff, while the beak is reddish brown. The other details of colour and pattern are sufficiently apparent in the illustration.

This gorgeous breeding-livery is, however, worn only during a portion of the year. For four months—namely, from June till September—the mandarin drake is a very ordinary-looking bird, clothed in a greyish dress, which in autumn gives place, by a second moult, to the breeding-livery, the latter lasting till the following summer. When the drake is in the non-breeding plumage, it is mainly distinguishable from the duck by its somewhat superior size. The colour of the female is chiefly grey, relieved with brown and white.

The distributional area of the mandarin duck includes northern China, Japan, and Amurland; but the species appears to be nowhere common, and in most districts is decidedly rare. In China these ducks, which are regarded as semi-sacred birds, are highly valued, and are kept in cages.

Unlike the majority of the duck tribe, both the mandarin and the summer duck are in the habit of perching on the branches of trees, and make their nests in holes either in the branches or trunk. In this perching habit they resemble the tree-ducks of the genus *Dendrocycna*. After lining the hole with a plentiful supply of feathers, down, and other soft material, the female lays a clutch of from seven to fifteen eggs. These she incubates herself, carefully covering them up with down every time she leaves the nest; and she likewise takes sole charge of the ducklings, from the time they are hatched till they are able to shift for themselves. When, as is frequently the case, the nest is situated over water, the ducklings throw themselves down into the water very soon after they are hatched, but in other instances they are carried in the beaks of their parents, as is the case with those of the ordinary wild duck when the nest is built above the level of the ground.

The food of these ducks, like that of many other members of the tribe, includes insects, worms, water-snails, young frogs, and other water-animals, together with the leaves and young shoots of various kinds of water-plants.

The mandarin duck is an active and untiring bird, moving quickly on the ground, and flying easily and rapidly into trees, where it flits from bough to bough with the ease and assurance of a wood-pigeon. It lives and breeds well in confinement, and may be seen in nearly all zoological gardens, as well as on many ornamental waters.

In the summer duck, which is a native of North America and Cuba, but occasionally wanders to Jamaica and the Bermudas, the colour of the upper-parts of the male in the breeding-dress is chiefly glossy green, with the cheeks purple, and black patches on the neck, and white stripes on both the head and neck. The wing-coverts are partly blue, the flanks are brown, black, and white, while the breast is chestnut spotted with white, and the rest of the under-parts white; the beak being a mixture of black, white, purplish, yellow, and scarlet, and the feet yellow.

THE HERON

(*Ardea cinerea*)

THE heron, or common or grey heron, as it is frequently called in order to distinguish it from its relatives, has fortunately not shared the fate of the bittern, and is still more or less common in many parts of Great Britain, where it may be seen at all hours of the day standing mid-leg deep in some stream patiently awaiting the next passing fish. Its gregarious habits and the protection accorded to many of the ancient heronries in various parts of the country doubtless account for the survival among us of this handsome, albeit thin and ill-favoured, grey bird. Nevertheless, on account of its fish-eating habits, the heron has many enemies, and is relentlessly persecuted in certain districts, especially by those connected with fisheries. Recently, however, efforts have been made to check this persecution; and in East Sussex, where there are no trout-hatcheries, these birds are protected throughout the year.

Although, as just mentioned, herons may be seen fishing at all hours of the day, they are chiefly nocturnal birds, and thus the very opposite of the kingfisher, which always captures its prey by daylight. When a heron sees a fish within reach, it strikes with unerring aim like a flash of lightning, and usually seizes its victim crosswise in its spear-like beak. In the case of larger fish, it is stated, however, to attack them by stabbing in the back—a mode of attack from which such fish, although mortally wounded, generally manage to escape. Of those fish which are seized crosswise in the beak, the larger ones are first beaten to death, after which they are swallowed head-foremost. When a successful lunge has been made, the heron resumes its motionless, watchful pose, confident that, although scared away for a time, the fish will soon return.

As the heron is a most voracious bird, consuming, it is affirmed, fully its own weight of food in a day, and as this food consists chiefly of fish, anglers, it must be confessed, have some excuse for the detestation with which they regard the species. Nevertheless, the heron does some good, as it also consumes a number of snakes and frogs, as well as water-rats. In addition to the above, herons also eat river-mussels, insects, worms, and probably also young birds.

The geographical distribution of the heron is very extensive, comprising most of the countries of the Old World, although the bird visits some of these only during certain parts of the year. In the north of Europe, for example, herons are migratory, travelling southwards to Africa in parties of as many as fifty individuals in October, and not returning till the following March or April. All streams and pieces of water in the neighbourhood of forests, or at least where a certain number of large trees are to be found, may serve as the fishing-resorts of the heron.

Herons build in large colonies, or heronries, which may contain from about fifteen to as many as four hundred nests. Formerly there were a great number of English heronries, especially in Lincolnshire, but many of these, like the well-known large one near Spalding, have been broken up. Many, however, still remain, and a new one has been recently established near Lewes. One of the largest English heronries was that of Bride, near Rye, in Sussex, which in 1860 contained as many as four hundred nests, although by 1880 there were barely a couple of hundred. Now, owing to the felling of some of the trees, this magnificent heronry has ceased to exist.

The nest is a large, rude structure of dry sticks and reeds, lined with hair, wool, and feathers, in which the female lays her three or four large green eggs. The young remain in the nest, or nesting-platform as it might well be called, until fully fledged, and are remarkably voracious. Putrefying fish cover the edge of the nest, as well as the ground below, and poison the air with their smell. The parents attend to their offspring for a few days after the latter leave the nest; but at the end of this period old and young part company.

If the nest be attacked, the parents suffer their eggs or young to be carried off, without doing more than opening their beaks and uttering mournful cries, although in many cases a single blow from the beak would suffice to slay the spoiler.

Heron-hawking was in former days a favourite sport of the nobles all over Europe. The falcon, usually the peregrine, always endeavoured to get above the heron, when, after the delivery of a successful attack, both birds fell together headlong to the ground. As a rule, the ornamental feathers—at one time highly esteemed—were plucked from the heron, which was then set at liberty.

On account of the damage it does to fisheries, the heron is even more persecuted on the Continent than in Great Britain; and is shot whenever an opportunity occurs, except in protected breeding-places.

THE GREAT SPOTTED WOODPECKER

(Dendrocopus major)

IT is not a little remarkable that such nearly allied birds as the great green woodpecker, or yaffle, and the two kinds of pied or spotted woodpeckers should present such remarkable diversity in the matter of colouring; the former being mainly olive-yellow with a red skull-cap in both sexes, while the other two are pied above and chiefly white below, with a red band at the back of the head of the cock alone. A third group is represented by the great black woodpecker, which is wholly sable, with an ivory-white beak. The difference in the matter of colouring between the green and the pied species is probably due to their different habits, the former being to a great extent a ground-bird, fond of frequenting lawns and meadows near woods for the purpose of digging up ants' nests, while the other two are almost completely arboreal. In the dappled shade cast by the leaves—especially those of pines—on the trunks of trees, these pied birds are comparatively inconspicuous; while among grass of moderate length the green woodpecker is absolutely invisible. It is further noteworthy that the pied species have most of the under surface of the body white, whereas in the green woodpecker the same aspect is grey. To a bird walking on grass a white under surface would certainly be no protection; but in bright sunshine on the trunk of a tree such a surface would undoubtedly tend to render the bird inconspicuous, as it would counteract the effect of the dark shade thrown by the body, in precisely the same manner as in the case of white-bellied quadrupeds.

Very curious is the fact that while, as already mentioned, the red band on the back of the pied species occurs in the adult only in the males, such a band is found in both sexes of the immature birds. This fact, coupled with the occurrence of a red head in both sexes of the green species, may be taken as an indication that red on the head was at one time a feature in all woodpeckers, but that for some reason it has been discarded in the females of the pied group.

Woodpeckers present some of the finest examples of the adaptation of bodily structure to be met with in the whole animal kingdom. The strong, conical beak is, for example, admirably suited for chiselling out, by repeated blows of the head, rotten wood in insect-infected trees, or prising off loose pieces of bark in order that the bird may be able to get at the insects and other creatures lurking beneath. Then, again, the short legs and the curious structure of the feet, with two toes turned forwards and the other backwards, enable these birds to obtain the most effective foothold on smooth, slippery bark. Lest, however, the feet should prove ineffectual, the bird is aided in climbing by its tail, the feathers of which have unusually strong quills, the tips of these being bare and shiny. When this tail is pressed firmly against the bark, the stiff tips of the feathers afford very considerable support to the ascending bird.

The climax in the way of special adaptation is presented, however, by the woodpecker's tongue, which, owing to the form and structure of the supporting bones, can be thrust out a long distance in advance of the tip of the beak, and is covered with a sticky secretion to which insects adhere. A similar structure obtains in the tongue of the wryneck; but there are certain foreign woodpeckers in which that organ is normal.

By means of its strong beak, the spotted woodpecker chisels out in the trunk of a tree, where the wood is more or less decayed, both a sleeping and a nesting hole. A circular entrance leads for some distance horizontally into the heart of the stem, after which the hole descends vertically for some way, and then expands into a large dome-shaped chamber, which serves as a receptacle for the clutch of three to eight white eggs, these being incubated by the male and female birds alternately. It is very generally believed that woodpeckers live entirely upon insects and other invertebrate animals; this, however, is a mistaken idea, for they likewise eat various kinds of seeds and berries, as well as nuts and walnuts, which they crack in nuthatch-fashion. The cry of the pied woodpecker is either a short and sharp "hi, hi," or a harsh and resounding "hæ, hæ"; but a more familiar sound is the tapping on the bark of the stem or larger boughs by the beak, in order that the bird may ascertain whether the wood beneath is sound or rotten. Very characteristic is the undulating flight of a woodpecker, the bird generally dropping suddenly near the end of its course, so as to alight only a short distance from the root of the selected tree, up the stem of which it then rapidly climbs.

The larger spotted woodpecker frequents woods with different kinds of trees; but its special favourites are pines, poplars, and willows.

THE SENEGAL PARROT

(Pæocephalus senegalus)

ALTHOUGH the name Senegal parrot, or rather *perroquet de Sénégal*, has been applied to several distinct members of the parrot tribe, it is now by general consent restricted to the gorgeously coloured species forming the subject of the accompanying illustration. In addition to this name, it has also the titles of orange-bellied long-winged parrot and black-headed parrot, the first of which is the more distinctive, although the second is preferable on account of its conciseness.

The black-headed parrot is a native of Senegambia and some of the countries of the West Coast properly so called, although the exact limits of its distributional area, which probably extend a considerable way into the heart of the continent, are still imperfectly known.

This parrot has been a well-known bird in Europe from very early days, as it was mentioned by Aloysius Cada Mosto so long ago as 1445, and was again referred to by the naturalist Brisson in the year 1760. Large numbers of these parrots are at times imported into Europe, especially to Havre and likewise to Liverpool. At the last-named port immense consignments used to be received now and then, but as these were for the most part young birds a very large proportion died soon after their arrival, especially when purchased singly and separated from their companions. Such young birds used to be sold at prices ranging upwards from five or six shillings; but tamed individuals are worth from twenty to thirty shillings each, while the few specimens that learn to talk fetch much higher prices.

If caught sufficiently young, these parrots make admirable cage-birds, as they are strikingly handsome, and fairly hardy. Occasionally they will lay in captivity, if provided with a suitable nesting-place. Sometimes they become very tame, although they are always nervous and excitable birds, uttering when alarmed a curious grating sound, and when thoroughly terrified giving vent to a shrill, whistling scream of fear. As a talker, the black-headed species bears, however, no comparison to the common grey parrot; and it has even been stated that the former is totally unable to learn to speak, although this is an error. The adult cock, which is rather larger and handsomer than his partner, has the head, cheek, and the upper portion of the throat brownish or blackish grey; the back, rump, and upper tail-coverts are glossy grass-green, while the wing-quills are olive greenish brown, the wing-coverts green with brown middles, and the shoulders, together with the lesser under wing-coverts, yellow. Those portions of the upper surface not already mentioned, together with the throat and the upper part of the breast and the whole of the tail-feathers, are bright grass-green. The remainder of the under surface is yellow, passing into bright orange-red on the middle of the breast and abdomen, the yellow likewise embracing the under tail-coverts. The beak is dark horny grey passing into blackish brown, with the soft "cere" at its base, like the patch of bare skin round each eye, blackish; while the feet are dark brown. Bright colour reappears in the iris of the eye, which varies from sulphur-yellow to dark brown, probably according to age.

In the hen the head is light brownish grey, the under surface uniformly yellow without any tinge of orange, while the under tail-coverts are yellowish green instead of yellow. In size the female may be compared with a small jackdaw.

As is the case with many other members of its tribe, very little is known with regard to Senegal parrots in a state of nature. They are stated, however, to associate in small parties of about half a dozen, and to take up their quarters, whenever possible, in the tops of the huge monkey-bread trees, where they reveal their presence by uttering piercing screams at the approach of an intruder on their domains. In taking wing, and also when settling after a flight, they are stated to be somewhat awkward, but when once started they fly as straight and as swift as arrows. Details are wanting with regard to their nesting-habits; but, when the young are strong enough to fly, the whole party takes to wandering about, and then frequently do much damage to the banana, rice, maize, and other crops. In captivity, at any rate, the males perform a kind of love-dance at the commencement of the breeding-season.

The Senegal parrot is the typical representative of a genus, with rather more than a dozen species, confined to Africa south of the Sahara. That genus belongs to a subfamily (*Pioninæ*), of which the more typical representatives, such as *Pionus* and *Chrysotis* (Amazon parrots), are South American.

THE GOLDFINCH

(Carduelis elegans)

PROPERLY speaking, the name "goldfinch" ought to be the designation of the canary, but it was doubtless given to the well-known British bird long before canaries were thought or heard of in England, and the former has, therefore, an indefeasible title. And, after all, if the canary be put aside, the designation is really very suitable to the goldfinch, referring as it does to the bright golden wing-bar which distinguishes both sexes from all other British birds. Taking into consideration the fact that both sexes share the brilliant plumage characteristic of the species, the goldfinch can lay claim to be the most brightly coloured perching-bird indigenous to the British Isles and north-western Europe generally.

To describe the colouring of such a well-known bird would be altogether superfluous on the present occasion, more especially as it is so excellently shown in the Plate. It may be remarked, however, that there is some amount of individual variation in this respect, and that the development of the red area on the head and of the white spots on all the tail-feathers is a feature of the adults alone; and it may be added that in regard to colouring females differ from the males chiefly by the smaller extent of the red area, which may contain black spots.

To one variation bird-fanciers have given the special name of "cheverel"; this rare sport, when fully developed, being characterised by the wholly white chin, and by the white patch on the cheek extending upwards so as to unite across the back of the head with its fellow of the opposite side. The brown patch on the breast is likewise replaced by white. Every kind of variation between a typical goldfinch and the so-called cheverel may be seen; and it is thus evident that the latter is merely a partially albinistic phase of the former.

There is an idea, doubtless unfounded, that the cheverel, or chevil, as it is sometimes called, mates better with the canary, and is likewise a superior songster; and it is to the latter notion that it owes its name, which is apparently derived from an old English word *chefle* or *chevelen*, signifying to talk idly, or chatter.

The distributional area of the goldfinch extends from the British Isles to western and central Siberia, beyond which it is replaced by the grey-headed goldfinch (*Carduelis caniceps*)—a bird with which the large eastern race of the European species will interbreed. Although the goldfinch is only a casual visitor to Scotland, in Scandinavia its breeding-area extends some five degrees farther north than that of its cousin, the brambling.

Owing to the estimation in which it is held as a cage-bird, the goldfinch was almost exterminated from most parts of England some years ago; but now that bird-nesting has been to a great extent stopped, the species is gradually recovering its numbers, and may often be seen on many thistle-covered commons to which it was long a stranger. In addition to Europe and western and central Asia, the goldfinch also inhabits northern Africa, while it has been introduced into New Zealand and Japan.

Its favourite haunts are open lands on the borders of woods, plantations, fields with trees, parks, and commons and other waste grounds. Thick forests it studiously avoids.

The food of this bird consists of seeds, more especially those of thistles and burdocks, as well as those of the birch and the alder. As a rule, it seeks those on the plants and trees themselves, and not on the ground; and in picking out the seeds from thistle-heads, it may be seen hanging head-downwards and in various other graceful attitudes on the stems. From the nature of its food, the goldfinch is, indeed, a most valuable bird both to the agriculturist and the gardener, on whom it confers additional benefits by disturbing insects which take up their quarters in its food-plants. It is, therefore, worthy of protection on two grounds—its utility and the beauty of its plumage, to say nothing of its song.

The nest is a beautifully made structure, nearly resembling that of the chaffinch, and generally built by the female alone, who is cheered in her task by the continuous song of the cock. It is frequently built in gardens, often at no great distance from the house, generally at a height of from fifteen to twenty feet above the ground, the most favoured situation being the fork of a bough, in which it is so well secured that it will retain its place even when the tree is felled. The four or five black-spotted bluish green eggs are laid by the female in May, and are hatched in thirteen or fourteen days. The young remain in the company of their parents for some time after they have left the nest. In many parts of the Continent goldfinches collect in the autumn in large flocks, which in winter break up into small parties of from ten to twenty birds.

THE RAZORBILL or AUK

(Alca torda)

THE razorbill, or auk, which, in suitable localities, is one of the commonest of British sea-birds, has an interest all its own from the circumstance that it is the nearest living relative of the now extinct great auk, these two species being, in fact, the sole members of the genus *Alca*. Both these birds present a considerable superficial resemblance to the penguins of the Southern Hemisphere; and it seems to be due to this resemblance that the latter owe their name, for there appears to be little doubt that the great auk was the true and original penguin, or pinguin, and that the birds we now know by that name were so called by the old voyagers on account of their likeness to the former species. Such resemblance as exists between the two groups is, however, merely of the most superficial kind, auks being strong fliers, with feet of normal structure, whereas the wings of penguins serve the purpose of paddles, and the bones of their feet are quite unlike those of all other birds.

Auks, in fact, appear to be near relatives of the gulls and terns, which have assumed, in accordance with their mode of life, a partially upright position of body. For these birds, in common with guillemots, very frequently breed on the narrow ledges of cliffs, where it is obvious that an upright posture affords them greater facilities for movement and at the same time economises space. In accordance with this habit, razorbills, in common with other members of the auk tribe, lay pear-shaped eggs, which cannot well roll off the bare ledges of rock on which they are often laid. As a rule, each female deposits only one, relatively large, egg; while no female incubates more than a pair of these eggs at the same time. Sometimes, in place of a bare ledge, the egg is laid in a hollow in the rock, or, where the soil is of a suitable nature, in a hole excavated by the parent bird.

All members of the auk tribe are inhabitants of the cooler portions of the Northern Hemisphere; their place in the corresponding southern latitudes being taken by the aforesaid penguins.

In addition to its remarkable bodily shape, the adult razorbill is easily recognised by the great lateral compression and subterminal expansion of the beak, from which the bird derives its ordinary vernacular name; as well as by the deep groovings and wide band on the sides of this appendage. The curved white stripe running from just in front of the eye to the root of the beak is another distinctive feature of the species; and this, too, in a more or less distinct form, in birds of all ages, whereas in the young the groovings and white markings on the beak are wanting. Considerable difference exists between the summer and winter plumages of the adult birds. In summer the head and neck are sooty brown, while the back and wings are black with a greenish gloss; the beak and the rest of the under-parts, together with a narrow wing-bar, being white. On the other hand, the winter plumage of the adults, like that of the young in summer, is browner above, while the sides of the head and the fore portion of the neck are of the same snowy white as the under-parts of the body.

The razorbill inhabits the coasts on both sides of the North Atlantic, breeding far up on the Norwegian coast, on those of Iceland and the Faröe Islands, and on the opposite side of the ocean on the shores of Newfoundland, Labrador, and Greenland; latitude 70° about marking its northward breeding range on the American side, while in Europe the limit is about one degree less. Eastwards the range extends to Jan Mayen Island; while the southward boundary of the breeding area in Europe appears to be formed by the Brittany coasts. In great Britain these birds breed, in suitable localities, all round the coasts, inclusive of those of the Shetland Islands.

These limits do not, however, by any means indicate the whole range of the species, for in winter these birds visit the Mediterranean, and occasionally the Canaries.

Throughout the year razorbills associate in large flocks, although in autumn the numbers of these colonies are diminished, apparently by a portion of the birds going out to sea. Bempton Cliffs, on the Yorkshire coast, form one of their favourite breeding-places, where the birds congregate in thousands, in company with guillemots, and yield a large harvest of eggs. The laying season commences about the middle of May, but is at its height some days later; while eggs and young may be found together till late in June. By the end of July the birds have for the most part finished their breeding season, and by the first week in August nearly all have forsaken the cliffs for the sea, which is their true home, and on which they often pass the night.